A HOUSE OF GOOD PROPORTION

Images of Women in Literature

EDITED BY

MICHELE MURRAY

Simon and Schuster
NEW YORK

SBN 671–21471–3 Casebound edition
SBN 671–21472–1 Touchstone paperback edition
Library of Congress Catalog Card Number: 72–93509
Designed by Irving Perkins
Manufactured in the United States of America

Permission to reprint the material used in this book was given by copyright holders listed below. I am grateful to those individuals and publishers who made my job much easier by their assistance and especially wish to thank Dr. Simon Karlinsky of the University of California, Berkeley, and Dr. George L. Kline of Bryn Mawr College, for generously providing the letter and the poem by Marina Cvetaeva. Also Catherine de Vinck for her poem and Mrs. Robert Linscott for the two Emily Dickinson poems from a collection edited by her late husband.

"Nightwatch," from *Nightwood*, by Djuna Barnes: First printed by Faber & Faber, London, 1936; American publication by Harcourt, Brace, 1937; all rights reserved. Reprinted through the courtesy of Djuna Barnes.

"Big Volodya and Little Volodya," by Anton Chekhov: From *The Image of Chekhov*, edited and translated by Robert Payne, copyright © 1963 by Alfred A. Knopf, Inc. Reprinted by permission of the publisher.

Material by Colette: Excerpted from *My Mother's House and Sido*, copyright 1953 by Farrar, Straus & Young, Inc. Reprinted with the permission of the publisher.

"Love That I Bear," by H.D.: Reprinted by permission of Grove Press, Inc. Copyright © 1957 by Norman Holmes Pearson.

"Two Women," by Tibor Déry: Permission granted by Tibor Déry.

"Stepping Westward," by Denise Levertov: From *The Sorrow Dance*, copyright © 1966 by Denise Levertov Goodman. Reprinted by permission of New Directions Publishing Corporation

Excerpt from *Olivia*, by Olivia: Reprinted by permission of The Hogarth Press.

"I Stand Here Ironing," by Tillie Olsen: Copyright 1956 by the Pacific Coast Committee for the Humanities of the American Council of Learned Societies. From *Tell Me a Riddle*, by Tillie Olsen, a Seymour Lawrence Book/Delacorte Press. Reprinted by permission of the publisher. Originally published in *Pacific Spectator*. And a special thank you to the author.

"Kindness," by Sylvia Plath: From *Ariel*, by Sylvia Plath, copyright © 1963 by Ted Hughes. Reprinted by permission of Harper & Row, Publishers, Inc.

Excerpt from "Ghazals: Homage to Ghalib," by Adrienne Rich: From *Leaflets: Poems 1965–1968*, by Adrienne Rich, copyright © 1969 by W. W. Norton & Company, Inc. Reprinted by permission of the publisher.

"The Tunnel," by Dorothy Richardson: Chapter 1 from Volume 2 of *Pilgrimage*, by Dorothy Richardson. Reprinted by permission of Alfred A. Knopf, Inc., publishers.

The Threshold, by Dorothea Rutherford: The Estate of Dorothea Rutherford.

"Consorting with Angels," by Anne Sexton: From *Live or Die*, copyright © by Anne Sexton. Reprinted by permission of the publisher, Houghton Mifflin Company.

Excerpt from "Family Happiness," by Leo Tolstoy [translated by J. R. Duff]: From *The Kreutzer Sonata, The Devil and other Tales*, by Leo Tolstoy. Published by Oxford University Press.

"Sex," by Jean Valentine: From *Dream Barker and Other Poems*, by Jean Valentine, copyright © by Yale University. Reprinted by permission of Yale University Press.

"On Barbara's Shore," by Diane Wakoski: Reprinted by permission of Diane Wakoski.

"On the Balcony," by Patrick White: Permission granted by Patrick White.

Excerpt from *White Mule*, by William Carlos Williams: Copyright 1937 by New Directions. Reprinted by permission of New Directions Publishing Corporation.

"The Starless Air," by Donald Windham: From *The Warm Country*, published by Charles Scribner's Sons, copyright © 1960 by Noonday Press, Inc. Reprinted by permission of Curtis Brown, Ltd.

I would like to thank as well for advice, criticism, and comfort José and Catherine de Vinck, Marion Garmel and Carol Bergé. And most especially, for help above and beyond the call of duty, my editor, Alice Mayhew, and my husband, James, who had a hand in every part of the book—except for the faults, which are mine alone.

Contents

INDEPENDENT WOMEN

THE WIFE

THE MOTHER

SCENES FROM FAMILY LIFE

WOMEN LOST

THE OLD MAID

THE OLD WOMAN

THE UNATTAINABLE OTHER

Illumination can take place only through the subject knowing himself first. There is no object unless the subject participates in his own experience. One cannot explore a dimension unless the constellation of one's own consciousness is prepared to apprehend it.

—I. RICE PEREIRA, *The Nature of Space*

Introduction

When Freud threw up his hands with the cry "Woman, what does she want?," he spoke in truth not of Woman but of himself, testifying to the boundaries of human imagination. What *does* Woman want? No doubt what Man wants: the unattainable. And the unattainable is precisely what women were not allowed to seek. He for God alone—for questing and speculation, for divine dissatisfaction. She for God in him—tangible, finite, not to be pondered or sought for (not to tax her small brain), all speculation brought to an end by a human existence restricted to the bed and the breakfast table. Of all the deprivations visited upon Woman through the ages, this shuttering of mental horizons was—is—possibly the worst. Why should the question *Who am I?* be answered, by command of Church and state, only in and through the existence of a man? Why does the very same question without the man become shameful?

Yet . . . yet . . . to borrow from Shylock his lament, "if you prick us do we not bleed? if you tickle us, do we not laugh? if you poison us do we not die?" Reading for this book, reading for twenty years and more as a woman in love with literature, I could discover no evidence in imaginative writing that would separate the essential condition of Woman from that of Man. I've brooded over the whole untidy subject since childhood, trying to place myself in some acceptable relation to work that I loved, especially when I met masculine resistance as a graduate student and teacher of English. "Where are the great women writers?" was a frequent challenge. How is the achievement of an instructor of English enlarged by his sharing the sex of Shakespeare? I could never figure that out. Still, I wondered.

Was I condemned to second rank before I even began and by impersonal biology?

Of course, women characters in literature were something else, existing quite as fully as their male counterparts. The only consistent difficulty I remember lay in interpretation, since the overwhelming majority of teachers and critics were and are men and any woman who challenged an interpretation of a woman character was regarded as clearly hysterical and biased.* A good deal of richness has been lost in the past which will be gained in the present and future as women bring their insights to imaginative writing, drawing on the resources of their special histories and particular experiences to illuminate what was written for them too by men and by more women writers of talent than one might believe.

This book contains the work of a few of them, although I didn't limit my choices to women writers alone. I aimed for the inclusion of less familiar artists and for variety, as this is only a sampler of much more that is available. Everything in these pages (as well as much that had to be cut for space reasons) I think is first-class as *literature,* and that is why it is included. My introductory comments draw on these selections in general, on many others, on my gathering thoughts and experience. Working on this book has helped me to crystallize my own reactions, to speak from the self. One of the genuine breakthroughs of the entire Women's Lib movement has been to free women to do just that—speak from the self, without waiting to receive the accustomed benison of authority.

II

What does Woman want? To cast a shadow fully as long and as rich as the shadow cast by Man. She wants not to be the Other, arranged in her place by comparison with the One, he who makes the comparison.† And if the One is splendid, noble, free, speculative, in-

* See Wallace Fowlie's autobiography, *Pantomime* (Chicago: Henry Regnery, 1951), p. 146, for a delightful example.

† From Joyce Cary's *A Fearful Joy* (New York: Harper & Row, 1949), p. 191: ". . . it seems to him once more that women are a race apart in a world of which the fantastic difference is hidden only by their logical inability to detect and describe it, or by their natural bent toward dissimulation."

tellectual, honest, strong, and brave, what is left for the Other, against whose background his own relief stands boldly out? She must be delicate, ignoble, chained, practical, emotional, dishonest (but in a charming, feminine way), weak, and timid.

It is a peculiarity of the human condition that each of us is more fascinating to him/her self than another person can be, for we are privy to the ceaseless activity of our minds, see ourselves as infinitely fluid and changing, appreciate our unspoken wit and applaud our generous impulses before they become ambiguous acts, enjoy (even in misery) a micro-human comedy in self-contemplation. Compared to ourselves, any other person appears relatively static, moving by a series of jerks from posture to posture, rather like a metal toy in need of oiling. That I should find myself more interesting than I find the other does not mean that he *is* less so, but only that my absorption in myself is quite complete and in him necessarily partial.

Man, encouraged by education, custom, religion, philosophy, and art, looked into his own mind and found it a curious, delightful, and everchanging spectacle. The record of what he found is contained in the books of our civilization. What more could there be? Woman, naturally lacking those agreeable traits he discovered in himself, was well fitted to the lesser (but still honorable) station of life which was already hers. How splendid that the laws of God, nature, and Man sang so harmoniously together!

Nevertheless, in spite of the weight of family discipline, religious sanction, and social pressure, Woman also discovered *her* self to be fascinating. Through centuries of silence, she has refused to be what Man has taken her for. She only *seemed* to stand still for Man to make his proclamations about her essential nature, nor did she offer any rebuttal except in the privacy of her own mind. Over the years how hard to find the forbidden—a glimpse of a woman thinking what she would not think, reported without guilt on her part!

Hidden beneath the patriarchal robes of the new gods, women have kept the secrets of the old Great Mother; midwives, herbalists, washers of corpses and keeners for the dead, witches and sorceresses, they became, through their connection with birth and death (two realities stubbornly insubordinate to reason) a secret flowering, like the colorless asphodel of Hades. We speak of "a man's world," but what is a woman's world? For most of us it has been a world of birth and death, of food and love, of comfort and blood—a very basic world, so essential to human existence that art cannot do without it. Hence, even when women do not exist in the public world or go beyond the walls of their homes, they are found in literature, often in central roles.

In literature Woman attains equality with Man, for she is essential to the telling of most stories. It is curious to mark the discrepancy between Woman in the life of the novel and Woman in the life of actual society. Is it fear that has kept Man from noticing it?

There is a clear connection between the power of Woman over birth and death and Man's fear of her: he needs to separate himself from the instinctual life of blood, from silent mystery, by loudly asserting his domination over nature (to which, in the end, she must yield) and over Woman; he must proclaim the triumph of impersonal reason and logic (or of their facsimiles) over the "primitive" realities to which he has condemned her, making an enclosure out of nature's lattice work, at the same time imprisoning her and depriving himself of all her wealth. These are devious and complex feelings, running deep, inaccessible to conscious control and now growing less exigent as we move toward a way of life where birth loses its mystery and death occurs in a hospital bed. The same emotions are present as motive forces in great literature, where they serve as sources of tension. Men and women are seen to be balancing the claims of the body and the instinctive life against the demands of an imagination which carries them beyond such claims, without entirely denying them.

III

From the vivid denials of men who turned away from the little they saw and half comprehended, we know that the women who made for themselves a place in permitted activities were onto something potent and frightening. The witch burnings were only the most dramatic manifestation of such a fear.

The Greek tragedies likewise may be seen as a vivid expression of such male fears and of the revenge of the repressed. In society Woman lives cloistered and demeaned, but in works of the imagination she rises gloriously in her destructive power as Clytemnestra, Electra, and

Medea, or as some inhuman model of foreknowledge in Cassandra, before losing in the end to a death that sets all things well in the cosmic order. This is a reflection not of life but of men's deepest dreams of memory and myth. Except for the fragmentary splendors of Sappho, there is silence around the dreams of women, for all that we know they dreamed, being human; a silence lying almost untouched to this day, our words nibbling away at the edges of some remote and sleeping Antarctica.

Not much has been preserved of what women have written or wondered or thought or said or dreamed since the beginning of history; they were always an underclass, and the words of members of an underclass are absent from most chronicles. We look—almost in vain—for the true voice of a woman. What did they think of their husbands and fathers? Of their endless childbearing? What did they tell their daughters? How did they see themselves when they met together? Only in the last 250 years do we get anything more than the echo of many voices (Teresa of Ávila's autobiography waters a desert!). Centuries pass and we glimpse a veiled form disappearing through doorways, a mother gratefully remembered by a grown son, a queen or noblewoman noticed for her power. There are women in the Bible, but it is men who speak for them and of them, who bargain for them and pay the bride-price, who take concubines and ask to be warmed by young flesh, who bed them and betray them, whose tenderness and love for them is the gift of generosity.

We have some stunning pages: Judith's hymn of praise to the Lord, and Mary's Magnificat, "My soul proclaims the greatness of the Lord." Woman has a soul. There have been times and places where this has been called into question, sometimes with a stridency that suggests the power of what was denied. But it has never been forgotten.

In the *Tale of Genji,* by Lady Murasaki, and in Sei Shonagon's *Pillow Book* we see through the strangeness of time, culture, and language, the society of Heian Japan where the nobility of both sexes led lives of Byzantine restriction and complexity and where sexual division exhausts the categories of existence: even the language forms and the style of script depend on sex. There is a deliberate effort at dividing the complete range of experience into separate sexual worlds to enhance the exquisite pleasure of these worlds joining in the sexual act. The women are more restricted in action even so (though the men

are also bound by an elaborate code), but the fact that women wrote
both books suggests that among the nobility, at least, women had a
rare intellectual and emotional freedom.

These are exceptions to the general silence. It is tempting to imagine
the worst of the blank spots, for there is so little to resist our imputa-
tions, there is so much rage now, such a passion for revisionism, as if
by taking thought we could—and should be able to—rewrite both
nature and history! Such passion, for all its noble faith in the powers
of mind cut free from its moorings, is cousin to the technological passion which would recast the earth and its inhabitants into a mold
more amenable to manipulate by the machine and its servants—
the passion to attack, transform, destroy, and rebuild. What does it
tell us beyond the ferocity of its own need?

In *A Room of One's Own* Virginia Woolf underlines the paradox
that while Provençal troubadours exalted Woman as immeasurably
above Man, the unattainable Other, the Queen of Earth and Paradise, in real life she was the slave of any man who could force a ring
on her finger. Yet there is no reason to assume that these two extremes,
while known and lived, marked the whole of Woman's existence as
it was experienced *by the women themselves*. We know too little to
assume that much. Literature does not tell us straightforwardly about
the life of its times so much as it does about how the imagination
perceives the flow without boundaries, the events in time from which
meaning was drawn. Beyond the revelation of how others see us, the
bounties of art are how we see ourselves, the shape we give to events
by our endowment, not the tangible occurrence, dead at the moment
of its completion. And we must draw our inferences with care.

IV

Why bring in consideration of the imagination, for God's sake,
when the cry is for politics, for a political solution to oppression?
Because imagination is the womb of change, and it is poverty of
imagination that rules almost all of the current nonfiction writing on
this ragged and tattered subject: *Women*—a quick breakdown into
either/or, the building of tight walls and roof around a position, then
mistaking it for a full-sized house to live in. Either submissive or com-

petitive. Either a woman made in man's image or a pseudo-man. In a world so various! A "dainty" wife dresses her daughters in ruffles, stays out of writing checks and fixing doorknobs, and calls herself "feminine." Another woman pins her worth as "a human being" to a job that takes her out of the home. A man glues his "masculinity" to a certain job and not helping with the dishes. All of these people, and their swarming kindred, have thinned their imaginative vision of themselves down to Giacometti size; then they starve themselves to fit into the pencil space they have allowed.

Since her appearance in Genesis, Woman has been a prisoner of the imaginations of others, is seen trailing clouds of glory or dust or whatever has been flung at her, always as the object, not as the creator of her own self, fully fleshed-out in the primary imagination, which is, in Coleridge's celebrated definition, "the living power and prime agent of all human perception, and as a repetition in the finite mind of the eternal act of creation." Coleridge then distinguishes a secondary imagination (into which the mode of fiction largely fits), which he considers

> as an echo of the former, co-existing with the conscious will, yet still as identical with the primary in the *kind* of its agency, and differing only in *degree,* and in the mode of its operation. It dissolves, diffuses, dissipates, in order to re-create: or where this process is rendered impossible, yet still at all events it struggles to idealize and to unify. It is essentially *vital,* even as all objects (as objects) are essentially fixed and dead.

What is important here is not what the imagination *is* but what it *does,* how it enables us to create out of inert stuff a possibility that may grow into actuality. The imagination giveth life; we cannot live in a certain way, we cannot see ourselves as the people we wish to be, until we perceive the wished-for life and self in our imaginations. The imagination brings to fruition infinite possibilities, it annihilates nothing, it charges all our being with meaning too rich to be fully conveyed, yet sustaining us in the operations of our life. In contrast, failure of the imagination brings about the conditions so visible around us—boredom, anomie, ugliness of relationships, inability to *see* the other in his/her wholeness. A power of being is choked off at its roots. The imagination, says Coleridge, is *active,* it is not a passive reaction to the given but a new creation, a struggle against fixedness and death.

Literature is subversive because it taps that secondary and even primary imagination, it puts us in touch with the springs of life if we care to drink from them, and its power is not amenable to societies or governments. The suppression of Flaubert's *Madame Bovary* is a classic instance of a social order's attempts to protect itself against a book clearly understood to be subversive, although the court did not point to the source of infection, which was not sexual immorality but Flaubert's uncannily exact rendering of "the woman problem."

Woman begins to be problematical to herself and others when her consciousness develops to a point where she sees herself as entitled to an individual life rather than to an impersonal predestination. At the time of publication of *Madame Bovary* (1857), there were already women's movements and manifestos, a visible rustling of the doves in the cotes which disturbed the male guardians of social order. The book itself dramatizes with painful clarity what happens when Emma's consciousness of herself as an individual is neither attached to any possibility for action nor allowed to develop by meeting a single human being who comprehends what is happening to her. Thrown back upon rootless fantasies, she becomes a power of destruction, lost in the confusion between a reality which has no place for her and a dream world which founders finally on *money,* that most real element and a key factor in the subjection of women. The quality of Flaubert's understanding, apparent in the measure of his insight and objective sympathy, made *Madame Bovary* a truly revolutionary book, and therefore ripe for suppression. Politics itself recognizes its subordination to the active principle of imagination.

The nineteenth century is the point in time when Woman becomes problematical on a large scale. Until then the number of human beings of both sexes not bound to the wheel of necessity is minimal. When a culture is at a subsistence level, the labor of all is vital for survival, and no matter what the law may say, the requirements of life will insure in fact a reasonable equality. Which is to say that women suffered and men did, too. My grandmother, a woman of spirit and intelligence, was imprisoned by unceasing poverty, worked hard until childbirth and immediately thereafter, raised a family and ran a candy store; my grandfather, who stood for twelve, fourteen, or sixteen hours a day in sweatshops on his varicosed legs, was her equal in deprivation.

Suppose, though, my grandparents were no longer poor. Ah, then! With a little money his life would have changed a good deal, hers very little. Women were no more restricted in the nineteenth century than they had been earlier, yet these restrictions were felt more and had to be defended by men, they rose to greater consciousness because, in certain classes, accumulation of wealth led to leisure and educational opportunities which freed more and more men. The comparison began to hurt. On an unmechanized farm, both sexes have their places and work side by side; but with business, cities, and universities, which made men of the expanding middle class free but did not free women, the women began to be aware of deprivation.

To be sure, the middle-class woman in 1880 was only relatively deprived; she might look at the great mass of humanity, both male and female, and congratulate herself on a condition superior to theirs. More and more, though, she didn't. Nineteenth-century novels trace this bubbling of discontent and her growing awareness that she was dependent upon her husband, who kept her only as long as she pleased him and could legally turn her out into the street any time he chose.

Freud crystallized the ethos of the nineteenth century regarding Woman; his work may break free of his time in other places, but here he is rooted squarely in the psychological confusion resulting from the first concerted assertion by Woman of her personhood.

What *does* Woman want? What does anyone want? What do I want? Impossibilities. To want, for good reasons or bad, without rhyme or reason, to have little and want something, to have much and want more—how human! These impossibilities don't submit themselves to politics or rhetoric, to sociology or psychology, without continuing to play tricks on the mind that seeks order or a final statement. Out of this wanting in all its power and folly artists make stories and sometimes a handle for grasping the always extraordinary nature of human existence.

And so we come to a closer look at Woman in literature, which is not the same as Woman in history or women living now, or this woman who inhabits my own provisional life—unless the imagination acts on it and draws out of it more than the activity, more than the existence subject to time. The nineteenth-century women are dead, but Lucy Snowe and Dorothea Brooke and Clara Middleton and Nanda Brookenham, creatures of imagination, will die only when there are no longer any readers receptive to the *meanings* they both exemplify and transform.

v

"The Imagination may be compared to Adam's dream—he awoke and found it true."—John Keats.

Or, to put it another way, he awoke and found it Eve. And did he then insist on Eve embodying his dream, making a "truth" congruent with his secret, given vision? Whatever he wished, Eve had a mind of her own, and for that she was punished, not only by God, but also by the generations of men whose sense of what is owed to them by women has been affronted by Eve's independence and curiosity (Eve, the first scientist; Adam, the first bureaucrat disclaiming responsibility). Her daughters would be deprived of that opportunity by being closely watched, for *they* were the sinful creatures, the sensual half of mankind, seducing innocent men to their deaths— Odysseus' Sirens, the Lorelei, Thais, La Belle Dame sans Merci. They were defective men, the principle of destruction, consorts of Satan— Lord, how Adam did revenge himself upon Eve for insisting upon what she was, her naked self!

The imagination, it appears, abhors a vacuum, is most at home with baroque contours, forbears to distinguish closely between day world and dream world. This is one of Woman's errors, not seeing how much she was a victim of imagination; she has seldom asserted her being on those grounds, has not been able to believe in the indelibility of such an evasive force.

Literature, that fragmentary record of imagination, exists where words border on wordless space at the boundary between day and dream. Pointing to the darker place, itself stands in the doorway. And I emphasize this: woman in literature is the subject of these reflections, woman mediated by language and consciously shaped to make a *figura* from what is given, what is known, what is dreamt.

It is important to insist on this distinction, for I feel the pressure of unimagination in much of the current writing on women, the social-science bias of mind combining with anger to produce generalized assertions supported by selective quotations to make a picture of blocklike forms without chiaroscuro. When there is such a multicolored richness! My own bias, shaped by immersion in poetry and fiction, is all for the individual case, the exception, the range of lights and shadows so much a part of the literary mode of perception. Not that the individual can't be used to argue the general—it can, I use it, who doesn't—but it remains at bottommost itself before it points to anything beyond itself.

In these times reportage has been said to usurp the place of fiction, and to do as well as, if not better than, the novel or story in conveying the tangled insights that lie at the heart of experience. Without defending—who could?—the great mass of novels ground out year by year, I do not share this opinion. Facts are no more than facts, as inert as commas on a page when lacking words to surround and sentences to decorate.

It is the shaping imagination which gives facts breath and heartbeat, and it is in art that this imagination is found at its fullest stretch. How hard to tap the resonance of that understanding in our country today, when the drift is all the other way! True it is that we do not have among us—now or at any time—many artists capable of transforming and crystallizing the chaos of life around us with the force required for handling jagged and broken reality (always jagged and broken, not only in our times, there is some vast San Andreas Fault underlying the surface of human existence) so that it yields its maximum meaning. Also true that when such artists do appear—especially when, as right now, they happen to be poets rather than prose writers—they are seldom noticed or read with awareness of what they are offering. Who, living on the San Andreas Fault, will welcome a seismologist as a neighbor?

No work of reportage, no matter how remarkable in its own terms, displaces the artist, for the imaginative voltage of nonfiction is, quite simply, lower. It may tell us about more people, more places, more things; it tells us *less* about them, however, and less about ourselves. We grow accustomed to this lower voltage, it is easier to live with, there is more diffusion, less dependence on the deep symbol-making power that touches us beyond consciousness; more is spelled out, our apprehension can be satisfied with an appearance of hard work. Small wonder that the added jolt of electricity from a work of art causes shock or an instinctive withdrawal.

Especially in a culture so wedded to the illusion that fact confers knowledge, that fact confers power, that fact exhausts the categories of experience, the temptation to make fact out of assertion and inference, and then to make fact supreme, is difficult to counter or to resist.

When I read about the wretched life, the miserable history, the maltreated psyches of Woman, I work my way into the theme to discover that its force comes—oh, so often!—from the author's own genuine unhappiness and the frustration of her friends, extended to embrace all classes, times, and cultures, as if the *fact* of this group's very real grief could naturally document so much more than itself with a minimal investment in imagination. Reading about the horrors of Woman's past, I find myself in a marvelously simplistic universe where all women had it rough and all men had it grand, just grand, sticking to the letter of the law, beggaring their wives, filling them with children, then discarding them like old shoes if childbirth didn't do the women in first. There is scant room for love, esteem, and a mutuality of relations in practice, even if not on paper. I read all of this and I wonder. Maybe it *was* like that everywhere and at all times. Or maybe not. Possibly the writer—at least part of the time —has mixed up the miseries (and, yes, the splendors, however infrequent) of the general human condition with those peculiar to one sex. Aren't there other stories, other views? I look for qualifications.

I look also for some recognition of those impossibilities whose scratch can never be itched away, which do weigh heavily on many women (yet are barely felt by others) and not lightly on many men, those impossibilities which are the source of so much joy and misery independent of law and custom, of place or expectation, of sex or position, those impossibilities which are the meat and drink of the artist.

When I find over and over again in the selections I have made for this book (and in poetry and prose not included) a prevailing note of melancholy and of the insufficiency at the heart of domestic life, I take it less for a plea that women abandon monogamy, farm their children out, and become doctors and lawyers, than for a rock-bottom recognition of the radical insufficiency of *any* finite arrangement to wholly satisfy longings which refuse to be bounded by the limits of human time and materiality.

The myth of unattainable, inexhaustible love has grown and swelled in Western culture until it fills every available space in the relations between men and women, whether explicitly brought forward or not, whether affirmed or denied: the myth of Tristan and Iseult, taken apart and pieced together in thousands of ways (see *Anna Karenina, Doctor Zhivago, The Man Without Qualities, Wuthering Heights* . . .), all essentially insisting that in death and death alone is the final possession, the true satisfaction of a passion which insists upon leaping beyond the allowable bounds, imaging in the doomed desire and pursuit of the whole the longing of men and women not only for each other but also for that love which moves the sun and other stars. Pressing on the characters of story after story are lights and shadows which they may not know about but walk through as we the readers walk through them in the pages of a book and also in our lives.

VI

Nourished on literature, it is difficult for me to accept technology or politics as more than ways of making this night at an uncomfortable inn a bit more comfortable, without sharing the illusions which feed, after all, a considerable portion of the apparatus of the modern state. If the writer is almost always cast in an adversary role, it is in part because of the sense which underlies his best efforts that life is fragile, intractable, and ultimately unsatisfying, even in his reconstructions, which try to do it over again but better, yet stop short at the edge of realization, no matter what. In literature women are not disadvantaged, even if they are scorned; their losses and failures and incompletions are only part of the general drift to loss and mortality that snags all men and the world made from action, ideas, words. All, finally, goes down to night, men have no more staying power than women, a jade cup or a gold brooch will last long beyond either of them.

Something else handicaps literary people (with some exceptions) for life in the modern state and for belief in the efficacy of large-scale political solutions: viewing the world through the lens of literature means that you can't be totalitarian without an awful lot of pushing and shoving. There is *this* poem, *that* story, a particular character, an individual exception to the general, the stubborn petunia in the onion patch to be accounted for—or not accounted for. Lives of artists as well as the lives of their creations cry out their uniqueness before any other word. As soon as I see "all women were . . ." or "all men did . . ." in word, argument, or inference, part of me withdraws skeptically and wonders, Did they, now? Every last one? And I'm lost to the point being made, a hopeless recruit in any marching army.

Not that I doubt the oppression or oppose lifting of burdens and erasure of injustices—having felt them myself, I testify to their very *physicalness*. Only, nothing is that simple, not even oppression, which requires a relationship of sorts, a complex connection which may open itself up to the patient watcher who has less faith in answers than in questions. When a woman says that she is satisfied with her traditional role, that she is content to be what her husband wishes, to follow him and serve her family, how to determine the authenticity of her words? We all wish to please (some, a little; others, much) and to act so that we see reflected in the eyes around us approval of our acts, of ourselves ("love me for myself alone, and not my golden hair"), so that we are validated by those with whom we have our deepest emotional connections. We care to do what is expected in our milieu— at least part of the time (and that we do not wish to do so either all of the time or none at all is the source of a lot of trouble)—and we forge our identity from catching glimpses of ourselves reflected back from the mirrors of all the faces we encounter in a day, in a lifetime, and from our responses to what we see, what we make of what we have.

Picking up the thread again, it is here that I locate the depths of Woman's difficulties (and Man's too, but less so, for he has set the conditions and defined the bounds of authenticity), in the limitations pressed upon her by every force of society as she attempts to work at that already painful task: *Who am I?* When Adam is asleep, am I only his dream? These provisional selves I test against fashion's dictates or my lover's desires, do they make a unity or only fragments whose broken glitter suggests the beauty of that wholeness now beyond reach even in dream light? Being always what the Other wants, have I lost forever the knowledge of—let alone the capacity for becoming—what

I am? Story after story, poem after poem, in these selections, reverberates to the sound of these questions, regardless of other streams flowing through them. "After such knowledge, what forgiveness?" When once Woman dreams her own self and wakes to find it true, is there no way out but flight like Nora or the continuation of the dream in death—Emma Bovary, Anna Karenina, Edna Pontellier?

Literature sets a starting point for us by showing us what has counted in the relations between men and women until now, what may continue to count; that is, its strength is that it deals in what *is* (not at all the same thing as portraying social conditions or fact), not in wishes, not even the author's wishes. Fashion suggests that by wishing and dieting we can change the surface and that, changing the surface, we can transform all that matters. Its immediate attractions are obvious.

Literature insists that the price of transformation is self-knowledge down to the bone, that the world is intractable, that the self is the most fragile of possibilities and the necessary condition for fullness, for growing beyond the reflection of the Other, for casting the long shadow. Why do so many women writers deal in the masquerades of wish fulfillment, play themselves and their characters false? Not all, not the most ambitious, perhaps, but enough of them so that we know instinctively (and flinch from) the two-edged praise of "woman writer"?

Years ago I read in Martin Buber's *The Knowledge of Man:*

> Man wishes to be confirmed in his being by man, and wishes to have a presence in the being of the other. The human person needs confirmation because man as man needs it . . .
> We may distinguish between two different types of human existence. The one proceeds from what one really is, the other from what one wishes to seem. In general, the two are found mixed together.

Over the years these words come back. How difficult for anyone to find his/her being at any time, how especially difficult for a woman so accustomed to having her seeming taken for this barely awakened being! So many men are incapable of seeing a woman as she is (even if she *be* a unity), they respond to an appearance that answers some dream or need of theirs, they give approval to a seeming that counterfeits being and continues to gain approval only by an increasing unfaithfulness to an authentic self. And as Buber writes, what one is and what one wishes to seem are found mixed together, so that when a woman receives all the tokens of approval for what she wishes to seem, what is the incentive to become what she is?

When I read those interminable discussions with Otto Rank, Henry Miller, and Lawrence Durrell that Anais Nin reports in her *Diary*, what struck me again and again was the way in which the men took their view of women and the reality of women to be the same thing, without even a chink of difference between them. In those pages we see Man's desire to consume the entire world with his ideas, to leave no room for the independence of any form of life not previously sanctified by belonging to his *idea* of it. They are examples of his inability to accept—on faith, if need be—that another's being is as fully real as one's own, that what Woman wants and needs is what Man wants and needs: her self, not as a grudgingly granted gift, but as the partner to his self.

Still, that is only half the equation. Buber says later on:

> The widespread tendency to live from the recurrent impression . . . originates . . . in men's dependence upon one another. It is no light thing to be confirmed in one's being by others, and seeming deceptively offers itself as a help in this. To yield to seeming is man's essential cowardice, to resist it is his essential courage. But this is not an inexorable state of affairs which is as it is and must so remain. One can struggle to come to oneself—that is, to come to confidence in being. One struggles, now more successfully, now less, but never in vain, even when one thinks he is defeated. One must at times pay dearly for life lived from the being; but it is never too dear.

Here there is no question of on one side oppression and on the other acquiescence, with no explanation for a millennial state of affairs beyond the old argument from Woman's innate inferiority, demolished without a better one to put in its place. Here there is a door opening on some understanding of the pleasures to be found in submission, of the metaphysical ease of living under the wing and in the charge of another who gives, after all, so much and demands in return so little, only the abandonment of a being which is painful to experience. It need not have been this way: so much space for Man and his ideas, such cramped quarters for Woman and what she wishes to say or think or be, nobody yet knows what that is, nor does she herself, this breathing space is so new.

Life lived from the being—it is this that moves us in the great women characters of literature, and it is at this level, perhaps, where the great recognitions of self will occur in the future to illuminate roles and relationships not yet dreamt of. So far literature has lifted the cloth from off a corner—the only corner introduced to a consistent light.

VII

Given my predispositions, my readings, the ambiguity of my own experiences, it is not surprising that I set out to assemble a book whose selections did not add up to make a point or even two or three. What I wanted was open-ended, one of those tests set to discover divergence or convergence, a fluid structure that did not determine the range of reactions by its very framework, like those elegant low couches that imprison us in one sitting position. I was looking for *something*, to be sure, for a created character that crystallized for me vague shapes or anticipations floating around in my head, for a character that carried my perceptions further on, attached them to a solid image, for a creation that gathered into itself all the legion of half-formed dreams and expectations that any of us brings to a page of print.

Naturally, looking in a close-fitted work for one part—depiction of women—changes the relationship of all the other parts of the work to each other and to the reader. A lot is being slighted, certain formal patternings go by the board, there is loss for any gain, different elements fade or loom large. To look at a work of literature this way, for a single aspect of its totality, is rather like perceiving a sculpture in the round from one angle only. It is flattened out, foreshortened, but the light on that angle may shine on some neglected spots.

Drawing on a lot of reading (back, back to the Bobbsey Twins) I remembered variety, separate emotional universes, some more congenial than others, no single note resounding, even in the limited range I arbitrarily set—American, English, some French and Russian writers and one superb story from Hungary—no points to be plucked out, like feathers from a goose, then stuffed into a feather pillow of

an argument for either misery or exaltation to lie on. Not that such
examples couldn't be found! Any chance collection would yield material to confirm any number of stereotypes (bargain basement) or arche-
types (designers' salon). Why not? The stereotyping mind eats proofs
for dinner even if peas are served, each subtly different from its kin.
There is such comfort in order, even if illusory!

The artist slides in and out of types; so do we all in life and mind.
Impossible to avoid living out, even in part, the great mythic roles
assigned to men and women, the extensions in imagination and time
of the primary sexual roles—which doesn't mean that we can't put
our own fingerprints on them. These primitive figures lend us their
shadows and enlarge our acts by the cultic memory of theirs. For the
artist, the type exists before he comes to it; part of the power of the
great characters in literature depends, in fact, on the reader having
a sense of what is *not* expressed, of the dimensions of a character that
already exist in the realm of types, the deep underground spaces that,
like a cellar to a house, support a structure, however complex. The
great writer might add so much that his creation will take on the
burden of the types for the future, so that new creations stand in *its*
shadow. But he/she doesn't begin *ex nihilo*.

I looked for variety. There is such a glitter all about us! Why reduce
it to one thing or the other? The past, the present, these completed
works, our continuing response which calls them forward from their
own sleep into our dream, all mingle in each singular perception,
radiant with possibilities! I can't analyze *how* some pages of prose or
a poem make an impact, leave a mark, but it happens without analysis, doors open at times with the slightest pressure, or perhaps only
after a shove, and sometimes they stay open so that, like Alice, we
glimpse the garden on the other side.

What didn't touch me that way I didn't use; I began and ended
with that: novels, stories, and poems that pleased me. I deliberately
avoided material easily available elsewhere, the well-known classics,
for example—so there is nothing by Jane Austen, no pages from
Wuthering Heights or *Jane Eyre* (but a chapter from the less familiar,
superior *Villette*), nothing from *Madame Bovary*. With regret, there
is nothing from Virginia Woolf, important as a writer for her depic-
tions of women, a writer with a narrow range, true, but with the
strength to tunnel deeply into it, and important as a woman who felt

the pressure of "a woman's role" on herself, fought it, fought free of it, yet at what a cost! But her best work is easily found and needs to be read complete.

Nothing here from that entire universe of a book read and reread when I was an adolescent so that, going back to that small and dusky library, I could reach up to the very top of the row of dark wood shelves in the fiction section and pull down its satisfactory green fatness—Sigrid Undset's *Kristin Lavransdatter,* the saga of the young girl, wife and mother, the mistress of sorrows, in a time and place—medieval Scandinavia—where women had simply not been *seen* before. None of her modern stories equalled the achievement of that historical monument but, reading them again, I found them moving and prescient, those stories and novels about dissatisfied wives caught up in sterile marriages and trying to work out their destinies in the narrow confines of provincial Norwegian society. Unfashionable today, to be sure, in their technique and style, yet worth reading for all of that, but none of them at their best in excerpts.

And that is important, for I chose sections from novels only when they made reasonable sense alone. No plays—that's another whole world. Works by women writers whenever possible, although not exclusively, except for the poems. As great a variety as I could find.

Indeed, I organized them to illustrate various stages in a woman's life so that the variety was evident. There are so many ways of being a woman! (How hard to see that, to escape from the either/or box . . . like midges, these tiny exclusive categories swarm at dusk.) These selections only suggest the beginning of what can be. They do not tell us what has been so much as they record different ways of apprehending without foreclosing other ways of making sense out of the world.

Nor do they tell us the way "real life" was in any direct fashion. Was there ever a family so well orchestrated as the Harlowes? How can a novel written in the 1860s—say *Adam Bede*—about a time forty years earlier tell us in the 1970s about the "real life" of either time? Those years are vanished, along with all who experienced them; what *we* know of them we glean from objects which have survived and words which have come forward into our lives bringing whatever evidence they can. We patch together clues and our expectations and make a past, to be sure, often taking that for the real thing, but it isn't (yet maybe is even more "real"), any more than a work of art is a report on the way things are.

Our sense of the way things are comes from our sense of who we are, and it is to this personal vision that literature adheres. What is it that we ask of this arrangement of words on a page that can—somehow—create perennial images? A story cannot live our days for us, a poem offers only itself, not packaged wisdom. The closeness of chaos to our most ordered moments has led me to cherish the order and pattern conferred by imagination upon what we call time and experience. In this case I have gathered up images of women which resonate still for me and, I hope, for others as well.

VIII

There are in my notes and comments depths and breadths un-measured, signs of my own failures of understanding and also of my life pieced out with interruptions, an example of the general condi-tions of "a woman's life"—desk in the dining room, cooking in the kitchen, telephone calls to break a phrase, laundry to be ironed, chil-dren all over the house (see the letter of Russian poet Marina Tsve-taeva on page 239). Two months have passed from the beginning of this essay to its ending, these words; things happen, seasons change, mending is done, the lilacs blossom, the checkbook balance nags at me, I read this book and that, put it down, mark my place, report cards are signed, people are invited for dinner—and all the while I try to keep steady a few ideas, and fail.

At times there is a frustration almost visible, a psychological smog. But there is another view, not only consolatory—an advantage that places literature at the very center of a life (my life), a current flow-ing to unite the continuing moments of a day. Willy-nilly there is no room for Literature as a subject that quickens only in a classroom, just as there is no self that reads and another self that moves through the rooms of a house, but only the single self unfolding from one action to another, from making beds to reading Tolstoy. So too there are only books which have their being in the context of life's other occurrences and shed their light—if it is light—on an open and wrinkled surface. Books are no shadow life, it is all one, a good flow, and not to be despised.

Ten years ago I taught at Catholic University and lived nearby in a new apartment built in a barren part of Washington, all little row houses and saplings running up and down the hilly streets, none of a city's delights, no shops or theaters, no libraries, no public gathering places, only these little houses and their owners mesmerized by their separate patches of lawn, their fenced gardens. The bus stop was by a piece of wasteland. From here I went each day to teach English literature, to participate, in however humble a fashion, in the great chain of culture that circled Oxford and Cambridge, the London theaters, Dr. Johnson's essays on Shakespeare, Hopkins' poem on Purcell and Purcell's music, theories of poetry from Longinus to Eliot, Dante's unmeasured world, the Gothic architecture so badly parodied in the university buildings, and so on. What was the connection between that spacious world and the raw, narrow one I came home to? Whether these worlds were even *aware* of each other was problematical. To say *the same sky over them,* was it to say mere words?

But *I* was aware of them, going back and forth, I was the connection, and any others who made that journey or similar journeys, for it is living people who transmit culture, joining together all the arcs of their days into a circle and joining all the circles into what we call a culture, a country, an era. What is not taken into the living flesh and bone, consumed and given out, is dead at birth, not culture but a ghastly parody. Thus, I deliberately confuse art and life—where is art to take root, to touch us, if not in the midst of ourselves, in the midst of our days? If it is not to have its effect on our lives, where will it go?

We need scholars, critics, and teachers, we are enriched by seeing as deeply as we can into the texts, but finally literature lives in the reading, not in the textbook. Here is a poem asserting its presence along with the lilacs stuffed into an empty bottle, today's mail, a child's painting, a loaf of bread. Why not? Perhaps out of the confusion and interruptions of my life these notes and selections can stake out their place in the jammed spaces of others' lives, both for what they are and for what they suggest.

A HOUSE OF GOOD PROPORTION

Venus—Aghia Sophia

BY CATHERINE DE VINCK

Above the waves
the lady stands in the pink shell,
innocent and sexual, flowing
in the lover's mind:
white opalescence, mother of all images,
discovered, denied.
Her voice low, she speaks
of ancient things, diagrams
of inner measures that shape
the world into a single unit,
a house of good proportion
built for the living.
Wisteria, iris, daffodil,
her language opens the garden gate.
Did you know in the past
the innumerable ways of seeds
and roots, the design of branches
spreading a filigree of newspun leaves
into an enameled sky?
Did you know the concordance
that links not only heaven and earth
but the most antique fragment of baked clay
to the very blood that pumps your heart
full of desire and dream?

In the center of your eye,
the lady is naked but modestly covered
by hair and hand; she is not to be taken
lightly; she brings a taste for excellence,

for recovery, more than healing:
strength; not the power to slay passion
but to orient the blinding risk
of love.
Is happiness controlled, pulled
by strings held
in some enormous hidden will?
Or is it born of water and flame,
of unbetrayed trust, of pain
that grinds flesh and bone
into a fine powder that the wind lifts?

The lady made of moon-foam, stepped
down from her sea-charriot
to warm the night, to nurse
the man-child with a tender urgency
lightyears removed from pleasure,
from play. "Be gentle," she says,
"learn that your inexhaustible yearning
cannot take final rest in me.
If you take suck
it is not only at my breast
but from the rose-beginning of the day,
from the open bud of creation,
from the dancing tip of life.
I am an icon of the master work,
and if you, enthralled, reach out
for the body of the poem, I will lead
you beyond the walled room
of the shell, unbinding your sight,
cutting the knots of raging hunger,
waking you
beyond the limit of my own meaning
to the universal heart of the fire."

Catherine de Vinck (1923–   ), *whose single published collection,*
A Time to Gather, *from a small press* (Alleluia Press, Allendale, New

Jersey) only hints at the excellence of her poetry, has produced a remarkable oeuvre, working in isolation and without hope of publication. As a religious poet she belongs next to George Herbert and the Traherne of *Centuries of Meditations. A native of Belgium, she learned English only when she came to the United States in 1948. Both the purity of her work and her use of Christian themes have worked against the acknowledgment she merits.*

The Little Girl

"She entered, as Venus from the sea, dripping." From the first conception of this book, through all the additions and subtractions, I kept to this as the opening sentence which would introduce these stories and poems of "stages on life's way." A doctor with a wide general practice in Rutherford, New Jersey, William Carlos Williams attended thousands of childbirths, which doesn't necessarily guarantee empathy, only blunt knowledge. However, Williams was also a writer sensitive to the poetry of the commonplace, and in this opening chapter of White Mule he caught, in his clean, sharp prose, one of those life experiences most difficult to convey because so smothered under sentimentality and imprecision. Steering clear of hymns of ecstasy, he gives childbirth—the mix of feelings, the physical exhaustion, the disgust, the relief, the ordinariness of blood, water, milk, baby shit, on which such a superstructure of emotions has been built. How good he is with the mother, Gurlie! A far cry this from the antiseptic fiction/film/television delivery rooms, with unwrinkled infants, twinkling moms, bumbling dads!

White Mule is the first novel of a trilogy which continues with In the Money and The Build-Up to tell the story of the family of Flossie Herman, the infant Venus of the first sentence, who grew up to marry William Carlos Williams. It is a novel full of the taste and flavor of the period from the 1890s to World War I, an analysis of social forces and of business life as well as a study of intricate family relationships, the whole told in a style whose simplicity is a product of the most complete art. A neglected masterpiece, overshadowed by Williams' influential and remarkable poetry, it will someday be seen as one of our great novels.

WILLIAM CARLOS WILLIAMS
(1 8 8 3 – 1 9 6 3)

White Mule

She entered, as Venus from the sea, dripping. The air enclosed her, she felt it all over her, touching, waking her. If Venus did not cry aloud after release from the pressures of that sea-womb, feeling the new and lighter flood springing in her chest, flinging out her arms— this one did. Screwing up her tiny smeared face, she let out three convulsive yells—and lay still.

Stop that crying, said Mrs. D, you should be glad to get outa that hole.

It's a girl. What? A girl. But I wanted a boy. Look again. It's a girl, Mam. No! Take it away. I don't want it. All this trouble for another girl.

What is it? said Joe, at the door. A little girl. That's too bad. Is it all right? Yes, a bit small though. That's all right then. Don't you think you'd better cover it up so it won't catch cold? Ah, you go on out of here now and let me manage, said Mrs. D. This appealed to him as proper so he went. Are you all right, Mama? Oh, leave me alone, what kind of a man are you? As he didn't exactly know what she meant he thought it better to close the door. So he did.

In prehistoric ooze it lay while Mrs. D wound the white twine about its pale blue stem with kindly clumsy knuckles and blunt fingers with black nails and with the wiped-off scissors from the cord at her waist, cut it—while it was twisting and flinging up its toes and fingers into the way—free.

Alone it lay upon its back on the bed, sagging down in the middle, by the smeared triple mountain of its mother's disgusted thighs and toppled belly.

The clotted rags were gathered. Struggling blindly against the squeezing touches of the puffing Mrs. D, it was lifted into a nice

woolen blanket and covered. It sucked its under lip and then let out
two more yells.

Ah, the little love. Hear it, Mam, it's trying to talk.

La, la, la, la, la, la, la! it said with its tongue—in the black softness
of the new pressures—and jerking up its hand, shoved its right thumb
into its eye, starting with surprise and pain and yelling and rolling in
its new agony. But finding the thumb again at random it sobbingly subsided into stillness.

Mrs. D lifted the cover and looked at it. It lay still. Her heart
stopped. It's dead! She shook the . . .

With a violent start the little arms and legs flew up into a tightened
knot, the face convulsed again—then as the nurse sighed, slowly the
tautened limbs relaxed. It did not seem to breathe.

And now if you're all right I'll wash the baby. All right, said the
new mother drowsily.

In that two-ridges lap with wind cut off at the bend of the neck it
lay, half dropping, regrasped—it was rubbed with warm oil that rested
in a saucer on the stove while Mrs. D with her feet on the step of
the oven rubbed and looked it all over, from the top of its head to the
shiny soles of its little feet.

About five pounds is my guess. You poor little mite, to come into a
world like this one. Roll over here and stop wriggling or you'll be on
the floor. Open your legs now till I rub some of this oil in there. You'll
open them glad enough one of these days—if you're not sorry for it.
So, in all of them creases. How it sticks. It's like lard. I wonder what
they have that on them for. It's a hard thing to be born a girl. There
you are now. Soon you'll be in your little bed and I wish I was the
same this minute.

She rubbed the oil under the armpits and carefully round the scrawny
folds of its little neck pushing the wobbly head back and front. In
behind the ears there was still that white grease of pre-birth. The
matted hair, larded to the head, on the brow it lay buttered heavily
while the whole back was caked with it, a yellow-white curd.

In the folds of the groin, the crotch where the genitals all bulging
and angry red seemed presages of some future growth, she rubbed
the warm oil, carefully—for she was a good woman—and thoroughly,
cleaning her fingers on her apron. She parted the little parts looking

and wondering at their smallness and perfection and shaking her head forebodingly.

The baby lay back at ease with closed eyes—lolling about as it was, lifted by a leg, an arm, and turned.

Mrs. D looked at the toes, counted them, admired the little perfect nails—and then taking each little hand, clenched tight at her approach, she smoothed it out and carefully anointed its small folds.

Into the little sleeping face she stared. The nose was flattened and askew, the mouth was still, the slits of the eyes were swollen closed— it seemed.

You're a homely little runt, God pardon you, she said—rubbing the spot in the top of the head. Better to leave that—I've heard you'd kill them if you pressed on that too hard. They say a bad nurse will stop a baby crying by pressing there—a cruel thing to do.

She looked again where further back upon the head a soft round lump was sticking up like a jockey cap askew. That'll all go down, she said to herself wisely because it was not the first baby Mrs. D had tended, nor the fifth nor the tenth nor the twentieth even.

She got out the wash boiler and put warm water in it. In that she carefully laid the new-born child. It half floated, half asleep— opening its eyes a moment then closing them and resting on Mrs. D's left hand, spread out behind its neck.

She soaped it thoroughly. The father came into the kitchen where they were and asked her if she thought he could have a cup of coffee before he left for work—or should he go and get it at the corner. He shouldn't have asked her—suddenly it flashed upon his mind. It's getting close to six o'clock, he said. How is it? Is it all right?

He leaned to look. The little thing opened its eyes, blinked and closed them in the flare of the kerosene oil lamp close by in the gilded bracket on the wall. Then it smiled a crooked little smile—or so it seemed to him.

It's the light that hurts its eyes, he thought, and taking a dish towel he hung it on the cord that ran across the kitchen so as to cast a shadow on the baby's face.

Hold it, said Mrs. D, getting up to fill the kettle.

He held it gingerly in his two hands, looking curiously, shyly at that ancient little face of a baby. He sat down, resting it on his knees, and covered its still wet body. That little female body. The baby rested. Squirming in the tender grip of his guarding hands, it sighed and opened its eyes wide.

He stared. The left eye was rolled deep in toward the nose; the other seemed to look straight at his own. There seemed to be a spot of blood upon it. He looked and a cold dread started through his arms. Cross eyed! Maybe blind. But as he looked—the eyes seemed straight. He was glad when Mrs. D relieved him—but he kept his peace. Somehow this bit of moving, unwelcome life had won him to itself forever. It was so ugly and so lost.

The pains he had seemed to feel in his own body while the child was being born, now relieved—it seemed almost as if it had been he that had been the mother. It was his baby girl. That's a funny feeling, he thought.

He merely shook his head.

Coffee was cooking on the back of the stove. The room was hot. He went into the front room. He looked through the crack of the door into their bedroom where she lay. Then he sat on the edge of the disheveled sofa where, in a blanket, he had slept that night—and waited. He was a good waiter. Almost time to go to work.

Mrs. D got the cornstarch from a box in the pantry. She had to hunt for it among a disarray of pots and cooking things and made a mental note to put some order into the place before she left. Ah, these women with good husbands, they have no sense at all. They should thank God and get to work.

Now she took the baby once more on her lap, unwrapped it where it lay and powdered the shriveling, gummy two-inch stem of the gummy cord, fished a roll of Canton flannel from the basket at her feet and putting one end upon the little pad of cotton on the baby's middle wrapped the binder round it tightly, round and round, pinning the end in place across the back. The child was hard there as a board now—but did not wake.

She looked and saw a red spot grow upon the fabric. Tie it again. Once more she unwapped the belly band. Out she took the stump of the cord and this time she wound it twenty times about with twine while the tiny creature heaved and vermiculated with joy at its relief from the too tight belly band.

Wrapping an end of cotton rag about her little finger, Mrs. D forced that in between the little lips and scrubbed those tender gums. The baby made a grimace and drew back from this assault, working its whole body to draw back.

Hold still, said Mrs. D, bruising the tiny mouth with sedulous care—until the mite began to cough and strain to vomit. She stopped at last.

Dried, diapered and dressed in elephantine clothes that hid it crinkily; stockinged, booted and capped, tied under the chin—now Mrs. D walked with her new creation from the sweaty kitchen into the double light of dawn and lamps, through the hallway to the front room where the father sat, to show him.

Where are you going? For a walk?, he said.

Look at it in its first clothes, she answered him.

Yes, he said, it looks fine. But he wondered why they put the cap and shoes on it.

Turning back again, Mrs. D held the baby in her left arm and with her right hand turned the knob and came once more into the smells of the birth chamber. There it was dark and the lamp burned low. The mother was asleep.

She put out the lamp, opened the inner shutters. There was a dim light in the room.

Waking with a start—What is it? the mother said. Where am I? Is it over? Is the baby here?

It is, said Mrs. D, and dressed and ready to be sucked. Are you flooding any?

Is it a boy? said the mother.

It's a girl, I told you before. You're half asleep.

Another girl. Agh, I don't want girls. Take it away and let me rest. God pardon you for saying that. Where is it? Let me see it, said the mother, sitting up so that her great breasts hung outside her undershirt. Lay down, said Mrs. D. I'm all right. I could get up and do a washing. Where is it?

She took the little thing and turned it around to look at it. Where is its face? Take off that cap. What are these shoes on for? She took them off with a jerk. You miserable scrawny little brat, she thought, and disgust and anger fought inside her chest, she was not one to cry —except in a fury.

The baby lay still, its mouth stinging from its scrub, its belly half strangled, its legs forced apart by the great diaper—and slept, grunting now and then.

Take it away and let me sleep. Look at your breasts, said Mrs. D. And with that they began to put the baby to the breast. It wouldn't wake.

The poor miserable thing, repeated the mother. This will fix it. It's its own mother's milk it needs to make a fine baby of it, said Mrs. D. Maybe it does, said the mother, but I don't believe it. You'll see, said Mrs. D.

As they forced the great nipple into its little mouth, the baby yawned. They waited. It slept again. They tried again. It squirmed its head away. Hold your breast back from its nose. They did.

Mrs. D squeezed the baby's cheeks together between her thumb and index finger. It drew back, opened its jaws and in they shoved the dripping nipple. The baby drew back. Then for a moment it sucked.

There she goes, said Mrs. D, and straightened up with a sigh, pressing her two hands against her hips and leaning back to ease the pain in her loins.

The mother stroked the silky hair, looked at the gently pulsing fontanelle, and holding her breast with the left hand to bring it to a point, straightened back upon the pillows and frowned.

The baby ceased to suck, squirming and twisting. The nipple lay idle in its mouth. It slept. Looking down, the mother noticed what had happened. It won't nurse, Mrs. D. Take it away. Mrs. D come here at once and take this thing, I'm in a dripping perspiration.

Mrs. D came. She insisted it should nurse. They tried. The baby waked with a start, gagging on the huge nipple. It pushed with its tongue. Mrs. D had it by the back of the neck pushing. She flattened out the nipple and pushed it in the mouth. Milk ran down the little throat, a watery kind of milk. The baby gagged purple and vomited.

Take it. Take it away. What's the matter with it? You're too rough with it.

If you hold it up properly, facing you and not away off at an angle as it— Mrs. D's professional pride was hurt. They tried again, earnestly, tense, uncomfortable, one cramped over where she sat with knees spread out, the other half kneeling, half on her elbows—till anger against the little rebellious spitting imp, anger and fatigue, overcame them.

Take it away, that's all, said the mother finally.

Reluctantly, red in the face, Mrs. D had no choice but to do what she was told. I'd like to spank it, she said, flicking its fingers with her own.

What! said the mother in such menacing tones that Mrs. D caught a fright and realized whom she was dealing with. She said no more.

But now, the baby began to rebel. First its face got red, its whole head suffused, it caught its breath and yelled in sobs and long shrill waves. It sobbed and forced its piercing little voice so small yet so disturbing in its penetrating puniness, mastering its whole surroundings till it seemed to madden them. It caught its breath and yelled in sobs and long shrill waves. It sobbed and squeezed its yell into their ears.

That's awful, said the mother, I can't have it in this room. I don't think it's any good. And she lay down upon her back exhausted.

Mrs. D with two red spots in her two cheeks and serious jaw and a headache took the yelling brat into the kitchen. Dose it up. What else?

She got the rancid castor oil and gave the baby some. It fought and spit. Letting it catch its breath, she fetched the fennel tea, already made upon the range, and sweetening it poured a portion into a bottle, sat down and rather roughly told the mite to take a drink. There, drat you. Sweet to unsweeten that unhappy belly. The baby sucked the fermentative warm stuff and liked it—and wet its diaper after.

Feeling the wet through her skirt and petticoat and drawers right on her thighs, Mrs. D leaped up and holding the thing out at arm's length, got fresh clothes and changed it.

Feeling the nice fresh diaper, cool and enticing, now the baby grew red all over. Its face swelled, suffused with color. Gripping its tiny strength together, it tightened its belly band even more.

The little devil, said Mrs. D, to wait till it's a new diaper on.

And with this final effort, the blessed little thing freed itself as best it could—and it did very well—of a quarter pound of tarrish, prenatal slime—some of which ran down one leg and got upon its stocking.

That's right, said Mrs. D.

The child in The Threshold *is just past the youngest ages at the start of that deep, mysterious, and almost unknown book; at the end, she is leaving childhood behind. She has begun to* understand. *Death has taken the little sister Irmgard, whom she has long resented, the beautiful and cold mother cries at night, the enormous adult world has shrunk in size. Throughout, we see this change only through the child's eyes, although the language, of course, is far more subtle and developed than what she could achieve. The combination of this disconnected viewpoint with a sophisticated syntax gives the book the quality of a dream remembered and elaborated upon. The following is a chapter not quite halfway through, a classical story of childhood loss unperceived by the adults,* The Olympians, *as Kenneth Grahame called them, and of a secret sorrow chipping away at the child's wonder and uncomprehending trust. Beneath the external differences—for this child is growing up long ago in a privileged family living in a Baltic city—there comes the recollection of our own losses, almost forgotten. How could we bear to remember them all?*

DOROTHEA RUTHERFORD

The Threshold

The toy-cupboard stands near the window. Between it and the window sill is a dark corner where the doll, Sidney, lives. It is certainly dark there, but it is also sheltered, and sometimes the sun shines in and points its long brilliant finger at the pallid Sidney lying in her white wooden bed, staring into space.

The low window sill is Liesbeth's playground. There she spreads out all her things, even her picture books, or she climbs onto it and sits leaning against a corner, sucking her thumb. There she is hidden behind a curtain and need not worry about being seen; and one can think much better while sucking one's thumb.

In front of the wide window, bars have been put up. Standing on the sill one can hold onto and look over them, since one does not want to see everything striped, especially when something exciting is happening, as, for instance, when the cab horse fell down and had to be unharnessed and a large crowd collected in the street waiting for it to burst (that, in any case, was what Maria said).

On that occasion, Liesbeth shared her window with Helga and Maria—who had suddenly become great friends, though normally they hardly exchanged a word. The misfortune of a third person often unites the bitterest enemies, even the superior with the nonsuperior.

"What's going to happen to the horse?" Liesbeth asked and pressed her chin so close to the bars that it hurt, in order not to miss anything.

"It will burst," declared Maria.

"But why? What do you mean, 'It will burst'?"

"Because it has eaten too much. That's what happens when you're greedy," Helga put in.

"Look what a fat belly it has," exclaimed Maria excitedly. "Watch it; it's getting bigger and bigger!"

"How dreadful! Do you think there'll be a loud bang and . . ." Beyond that Liesbeth could not think.

"Don't ask so many questions," interrupted Helga crossly, "and come away from the window!"

But Liesbeth paid no heed. She clung to Maria, who seemed to be much better informed.

"What will happen after the bang, Maria? Will everything inside the horse come out?"

"Of course, what else?"

Liesbeth shuddered. How awful! She would rather not see it. Then she asked, terrified of Maria's possible confirmation: "Is it the same with people?"

"Yes," Maria replied absently, being at the moment more interested in the crowd outside than in Liesbeth's thirst for knowledge. Two policemen had by now joined the throng; a pair of inquisitive small schoolboys held their ground and tried to press forward. Maria caught sight of someone she knew in the crowd.

"The same with children, Maria?"

"Of course, if they eat too much." Liesbeth started to feel her stomach. Wasn't it already beginning to swell? No, not yet. But it could happen, any day. . . . She must be very careful.

Meanwhile the horse, after much combined effort and many encouraging cries on the part of the driver and the policemen, was brought back to its feet and the performance was over. Liesbeth was sorry that nothing had happened. That was often the case when you were excited about something: it all fizzled out. But secretly she was glad that she did not have to watch the horse burst. . . .

The toy-cupboard had belonged to her father when he was a little boy. Now it contained Liesbeth's and Irmgard's books, playthings and other treasures. Each had a shelf of her own. Down below, where there had been two drawers before, there were now two deep compartments, which was, in fact, better, since they held much more. There one could hide all one's most secret treasures, for nobody would have the energy to begin clearing out the compartment belonging to Liesbeth. Last Christmas she hid in it the pig's tail which she had begged from Maria. Such a lovely, pink tail it was! You couldn't use it for sausages or anything good to eat, so Maria had allowed herself to be persuaded, and Liesbeth had quickly secreted it in her compartment. But that was a long time ago. Today something much more valuable lives there . . . in quite a different class from the pig's tail. Carefully covered up with many-colored silk and woolen rags, bits of string, corks, empty matchboxes, and an infinite number of other treasures, her "cobblestone" lives in the deepest, darkest corner of the compartment.

When you have secretly and hopelessly longed for something and then suddenly got it, naturally you want to keep it in a place unknown to everyone where it is sheltered from profane eyes. Merely the thought that it is there gives Liesbeth a thrill every time she passes the cupboard, a thrill like a flash of lightning in which joy and pain are one.

A cobblestone is square and dark brown and covered with a thin layer of sugar icing. It smells of honey and spice, and when you take a bite it is brittle and spongy at the same time, for it sticks pleasantly and thickly to the gums and teeth. But it isn't too sweet, that's what's so good about it. And, besides, it isn't only a cake, but Christmas itself: if you hold it in your hand you can smell the faint fragrance of burning candles and warm fir branches, you can see the white, glittering starry night, you can hear the angels singing in triumphant chorus, "Holy Night . . ."

Therefore a cobblestone is something extraordinary. It is magic. It is, in itself, a whole Christmas tale.

In Herr Stude's sweetshop in the High Street such a cobblestone is almost unnoticeable amongst all the other splendors, but it is the only thing Liesbeth craves. Perhaps she is too modest to hang her hopes on a piece of delicious marzipan, or a sausage sandwich, or a piece of Swiss cheese with the tiny holes in it.

There are so many splendors in Herr Stude's they would stun the imagination, for Stude is not simply a confectioner but an artist. True, he doesn't model in marble or clay, but conjures his masterpieces out of sugar, fresh eggs, almonds, nuts, cream and vanilla. There are plates covered with fried eggs, called also ox-eyes, brown fried *rissoles* with potatoes and green peas, ham sandwiches and anchovies. All one has to do is to walk in and say, "Dear Herr Stude, I would like to send my friend a crisp roast goose for her birthday," and it is as good as done. Nothing is impossible for Herr Stude, therefore Liesbeth is convinced that he and God must stand in close relationship.

When he presides, tall and serious, behind the counter and Liesbeth looks up at him and sees his long face with the well-trimmed graying beard and the glittering spectacles, she feels the same awe as before Father Christmas. Herr Stude hardly ever laughs; he is always serious and composed. He has obviously more important things to think about than silly little girls.

Liesbeth is tall for her age, so that she can see, just above the counter, the rows of little cakes, which no one in the world, as everybody knows, can produce as well as Herr Stude. There are the meringues—small, sweet kisses of white of egg, vanilla and whipped cream; there are the "Negroes' heads" covered with chocolate and filled with light yellow cream; there are the Alexander cakes that melt on your tongue when you bite into them—long sticks of pastry with a crust of browned sugar and the whipped cream running out at the bottom no matter how careful you are, so that you must hold your hand underneath in order not to waste any of the deliciousness and afterwards lick your fingers. How is one to enumerate and describe all the sweet fantasies called "cakes"? It would need a book.

On the left, carefully arranged under glass, lie the candies: chocolates and caramels, with a crunchy sugar coating on the outside and a fascinating filling on the inside that can stretch and stretch, pink and bright yellow. No, you would have to write a volume to do justice to them all.

When you have a cold, Aunt Agatha brings you a little box of barley sugar. The name alone, *Georg Stude,* painted in pretty script letters on the box, fills your heart with delight. The contents are so good that you can only wish for a cold since otherwise you don't get any.

The cake counter must never be empty. As soon as there is a gap, a hidden door opens in the back wall and a kitchen boy carries in on his shoulder a large tray laden with cakes so that the young lady behind the counter may immediately fill up the empty places with a pair of silver tongs. Just like "Table, set yourself," in the fairy story!

There are also matchboxes containing peppermint matches with chocolate tops, mushrooms in straw baskets, apples and pears, half-peeled oranges made out of the choicest marzipan and so lifelike that you can have a lot of fun mixing them with real ones in a bowl and offering them to visitors. For tourists who want a souvenir there are views of the town made of white marzipan in little flat boxes. You can have "Fat Margaret" and "Tall Herman" with the Castle, or the whole town with its towers and gables seen from the sea.

No wonder that the door at Stude's can hardly gather breath, that the little bell rings from early morning till late at night, especially in winter. With every silvery *ping,* a fresh draft keeps streaming into the chocolate-and-vanilla-scented atmosphere. Big and small, old and young, poor and rich come in and out, for everyone who possibly can saves a little, to forget for a while the vexations and troubles of life, even if they can only run to a "Negro's head."

The lady at the cashbox must lead a wonderful life. You can see it from the friendly smile she has for everyone. A nickel chest stands beside her and a little flame burns beneath it. One need only pull out a drawer to take out a warm meat pie. When the weather is so cold that drops freeze on your nose and the horses have icy white beards, this has a reviving effect on your spirits and your tummy.

Yet of all these splendors Liesbeth longs only for a cobblestone. She seldom goes to Stude's, so that it is a great occasion when Mother takes her there.

Liesbeth summons all the courage she possesses and pulls her mother by the arm. "What do you want?" the latter asks impatiently, and does not even look at Liesbeth, because she is discussing a fruit tart with Herr Stude.

Liesbeth feels hot and wet in her padded coat, fur collar and cap. If only she can muster enough courage! "Please, please," she whispers and tugs more violently at her mother's sleeve.

"Well, what do you want?" (Oh, if only she doesn't get angry!) "Speak up!"

How can one say it out loud! Can one shout out a prayer that lies buried at the bottom of one's heart? And in the end Mother will certainly be angry and then the whole day will be spoiled, for Liesbeth loves and admires her mother with every fiber of her being. She looks up with imploring eyes. If only Mother would guess! Must one always put things into words?

Liesbeth does not know that there are very few people who don't have to have everything set out plainly and clearly. She is also unaware that words often mislead instead of explaining. She looks up imploringly and mumbles, "Please, can I have a cobblestone?" (Will the world come to an end?)

"Speak *up*," says her mother. "What's the matter with you?"

Once again she has to make the effort. In a voice that appears strange even to herself she speaks up as loud as she can—or does it only seem to her that she is shouting? Now they can all hear her begging, the old and the young and the big boys in school uniform before whom she is particularly shy. "Can I have a cobblestone?"

"No," says her mother very firmly. "Why should you? It isn't Christmas yet."

As they turn to go Liesbeth is perspiring with shame because she has been begging. Never, never will she do it again. How should she, small and stupid as she is, ever aspire to such splendor? The cobblestone, which a moment ago still seemed important, has become insignificant beside her shame. Her lower lip moves in and out, trembling. Her whole face refuses to obey her.

"No need to sulk," her mother says reproachfully. "Stop pouting! If you knew what you looked like!"

Ping-ping, says the door, more merrily than ever, as though it were poking fun at Liesbeth, and the cold air rushes at her like a ravenous dog. But she feels hot behind the eyes.

It is sad, of course, that she couldn't have her cobblestone, but much worse is that she begged for it. When you bare a secret dream and someone stretches out an importunate hand toward it, you shrivel to nothing. How can you explain such a thing? There are no words. . . .

They go through the passage, long and narrow as a corridor, where Liesbeth sees the beggars leaning as usual against the wall of the Church of the Holy Ghost. Every time she passes them she wants not to look, but her gaze is irresistibly drawn to them. There they sit on the cold stones, wrapped in colorless rags, their crutches and sticks beside them, exhibiting their sores and defects. It is horrible. Liesbeth's stomach turns inside out and she averts her face to avoid the foul smell that emanates from them, but her eyes are forced back to them despite herself. She is sorry for them, but they are so dirty and hold out their palms to the passers-by for coppers. They exhibit ulcers which they ought to keep hidden and use them for begging.

This time Liesbeth stops in front of them. "Can I give them something?"

"No, not in any circumstances," says her mother, dragging Liesbeth away. "One should not beg."

There it is, one should not beg!

Liesbeth barely sees the little stalls that cling to the wall like swallows' nests and usually give her so much pleasure. She does not even notice the apple women squatting on their wooden boxes, whose loud quarrels usually interest her so much. The hungry looks of the beggars seem to be pursuing her. . . . Has she not just been begging, too? The skinny sore-covered fingers seem to be touching her and she shudders. Never, never again will she think of a cobblestone. Yet at the bottom of her heart she believes in miracles. This belief stirs in her a bud that wants to burst on a branch because it believes in light, because it feels that the tree in which it slumbers concealed is not everything. And as long as there are miracles, even if perhaps not for her, all is well.

But you don't beg for miracles. They happen . . . and when you least expect them. When you don't ask for them.

And they happen even to her! How does Aunt Agatha know that you are still secretly longing for the cobblestone? Aunt Agatha is like no one else. She appears one day in her hat and coat and with the umbrella without which she never leaves the house, and which Liesbeth admires so much because it has a dog's head for a handle. She is nodding her head violently, a thing she does when she is excited or happy, and sometimes Liesbeth is afraid that her head may fall off. Today she has a mysterious parcel in her hand and her whole face is laughing. You notice that she's been out because she brings such a fresh clean breath of frost in with her. And a cobblestone for each child!

"You spoil them, Agatha," says Liesbeth's mother, but in a flash Liesbeth has the soft, sweet-smelling cake in her hand. How lovely it is! Like a bride under the thin veil of sugar! And it speaks to her with its sweet breath.

Liesbeth sits on the window sill in the nursery and holds it with both hands. She inhales its spicy, sweet aroma, runs a cautious finger over it. Shall she bite off a piece? Nibble at a corner? No, then it won't be a whole, intact cobblestone any longer. It will be a cripple!

But you can, of course, lick it very carefully and lightly, as though your tongue were the wing of a butterfly. That can't do it any harm. How smooth it is, yet it has such enchanting little bumps, there where the sugar was poured thickest.

Helga comes into the room. "What have you got there, Liesbeth?"

"Can't you see, Helga? A cobblestone." No, Helga must not look at it! She might ask for a piece. It must be hidden away. There is only one place in the whole house that belongs to Liesbeth—her "cave" in the toy-cupboard.

"Don't overeat," Helga says and leaves the room.

Don't the grownups think about anything but eating? Do they only see and hear what a thing is, and not what is most important about it: its scent, its music, its dream texture, delicate as pollen, resplendent as a thousand rainbows?

Liesbeth opens the toy-cupboard and a heap of dusty objects tumble out with a clatter. But that is not enough. She pushes both hands inside the dark cavity and pulls out heaps more, until a pile of debris lies before her. She fishes out a doll's leg, then a musical top. A bit bent, it's true, and the paint rubbed off in spots, but an old friend. Liesbeth wonders if it can still hum. The button you have to press is gone, but the deep beautiful notes surely still live in it and that is what matters. Now it keeps these notes for itself. That is why it is still a musical top, to be protected from Helga's furies of tidiness.

Liesbeth takes the cobblestone, which is lying on her apron, looks at it lovingly, kisses it, licks it tenderly and pushes it into the remotest corner of the "cave," where her eye cannot reach any more, only her hand. Then quickly, very quickly, before anyone comes, she thrusts in everything else. There is a great clatter and some of the things resist, but she pushes energetically with both hands and closes the cupboard door. She is as filled with a sense of satisfaction as a bottle with sparkling wine.

The whole day she goes about like one intoxicated. She does not walk, she floats, borne up by clouds. She has no legs, only wings, and she is so gentle and helpful that her mother fears she may be coming down with something serious.

The best part of it is when she forgets about the buried treasure for a few moments and then suddenly something joyous, exultant, knocks at her breast and reminds her that she owns a cobblestone. She turns her back to the cupboard as though she knew nothing about its secret and feels as in the song:

> *It knows and will not tell,*
> *How well with me, how well.*

Often she is drawn to the cupboard as to a magnet. But no, not yet . . . "For a rainy day," says Liesbeth, "for a gray, empty day. . . . Then I will give myself a treat."

Today is precisely such a day. An icy gray wind sweeps through the streets and a dark heavy sky lies upon the roofs. All the houses stand colorless and ill-tempered, people hurry as though they were fleeing to get back to their warm rooms, nobody stops to exchange a friendly word. This morning Father tapped the barometer, looked at the thermometer in the dining-room window and said: "It is going to snow."

But it isn't snowing, though the wind has gone down and the air is heavy with an expectant silence. The day is colorless and shadowless as a desert and everybody looks morose. Helga has wrapped an ugly woolen shawl round her shoulders, Mother has had words with Maria, and the latter walks about sniffling and has only short answers for Liesbeth. It is as though they were all bewitched.

Liesbeth alone is immune. Has she not a secret and the ability to transform this day into a real holiday simply by producing the cobblestone? Everybody will get a bit of it and it will work on them like a charm.

She goes to the toy-cupboard and kneels before it. After opening the door she sits there for a moment in rapture. Then she starts rummaging with both hands in the cavity. Now it must happen quickly. There is no longer any reason for delay.

A whole heap of things already lies in front of her, but she hasn't yet reached the cobblestone. It is hidden, of course, far back in the deepest darkness, where no one can see it. She spends no time stroking the doll's leg or talking with the silent top, so hard does impatience drive her now that the moment has come. She pulls the things out more and more quickly, making a great clatter. But the desired object remains concealed.

Now the cavity is empty and the great moment has come. Liesbeth thrusts both arms deep inside. How bottomless it is! It is as though it kept retreating. Her groping fingers feel nothing—nothing but rough walls and wooden moldings.

It's impossible! It must have hidden itself somewhere! Perhaps it has tumbled out with the other things and in her impatience she didn't notice it. She rummages with busy fingers in the debris lying half in her lap, half on the floor in front of her. There, under the shells which she picked up on the beach in the summer, glitters something brown! But it's only a bit of cloth.

A terrible thought grips her heart: What if it really isn't here?

No, that's impossible: she hasn't been near the compartment for a long time and it couldn't have got out by itself.

Again she gropes in the dark and again encounters nothing—nothing but wood. Her heart contracts and becomes a little dry hazelnut; it pounds right up into her throat and hurts.

"Helga," she stammers.

"What is it?" Helga asks and goes on dusting wearily at the other end of the room.

"Helga . . . my cobblestone . . ."

Helga shows no sign of hearing the anguish in Liesbeth's cry. She goes on dusting ill-humoredly, sunk in her own thoughts.

"Helga . . ."

"Well, what is it? What's all this about your cobblestone?"

"I can't find it."

Helga stops dusting for a moment and appears to be thinking.

"Do you mean that old piece of gingerbread in the cupboard?"

"My cobblestone, Helga!"

"You won't find it. Irmgard ate it yesterday."

Speechless, Liesbeth stares at her. It isn't possible. Her heart, which by now has turned into a lump of lead, swells, presses higher and higher, until she feels she is suffocating. She stares through a damp mist in which Helga appears and disappears and seems to multiply.

Helga goes on, while dusting, "It was no good any more, you know. Irmgard saw it and wanted it so much that I gave it to her."

How did Helga come to rummage in her cave? In her holy of holies?

"But it was *my* cobblestone."

"I've already told you that it was no longer any good. All dry and dusty. You haven't lost much, I promise you."

"But why did you give it to Irmgard if it was no good?"

"She wanted it so badly and she is only a little thing, and as you had forgotten all about it I wanted to get rid of it."

Liesbeth is unable to utter a word. She sits benumbed in front of her pile of toys and smells the dust from the cavity.

Helga comes nearer. "Put your rubbish back into the cupboard. I want to dust here."

Very slowly, without a word, Liesbeth puts her things back into the cupboard, one by one. The doll's leg is really very dirty, the shells have lost their rainbow luster . . . how is it that she hadn't noticed it before? They are gray, like the house opposite, like Helga's woolen shawl. The top is all battered, the silk bits of stuff destined for a dress for Sidney are silly rags which would not be sufficient to cover her belly, and who ever made her believe that they were pretty?

At last everything is in. It took so long! Liesbeth closes the cupboard door and slowly leaves the room. Where to? It makes no difference. Her legs feel as though made of cotton wool and lead her into a spidery gray, hopeless day that stretches endlessly before her.

She steals into the sitting room. Mother is there, sitting at the piano, singing. She sings like a lark. The notes play round her bright and clear like drops of spray from a fountain. She is rapt in her song. Irmgard sits in an armchair with her legs drawn up under her and is also absorbed in the music.

Liesbeth edges close to her mother. She doesn't wish to speak, to see anyone; she only seeks comfort in her mother's warmth, for it is cold and she is at the mercy of all the winds. She leans with her back to the wall and sucks her thumb. Behind her a sparrow pecks at the window-pane. Then it flies away and doubtless chirrups louder than all these others quarreling over the horse dung. They have no Helga to steal away their oats.

Mother has finished her song and turns round. "Do you want anything?" she asks absently. "What's the matter?"

Liesbeth pulls her thumb out of her mouth and wipes it on her apron because it is so wet, but makes no sound.

"Well! Out with it! It's no good sulking."

"My cobblestone," murmurs Liesbeth to herself and bends her head even lower.

"Lift your head and speak louder—where are your manners? What's this about your cobblestone?"

Has she got to shout out all her misery so that it happens all over again in words? She would burst into tears, and Mother doesn't like tears at all.

"Out with it. . . . Yes, your cobblestone. . . . Go on!"

"Helga gave it to Irmi."

"And that's why you're sulking? But why didn't you eat it long ago? Aren't you ashamed of yourself? Fancy being so mean!" Mother seems quite incensed. "You grudge an old piece of gingerbread to your little sister. No, tears won't help. Go and wash your face and look a little more cheerful."

Covered with shame, Liesbeth slinks from the room. The nursery is empty; Helga has finished her work and gone to the kitchen to fetch a cup of coffee for herself. There is the familiar window sill. Liesbeth climbs onto it and presses herself deep in the corner. She feels abandoned by God and man.

Outside on the gray pavement a bitter wind is blowing. Except for a lady looking in at the window of the shoeshop, the street is empty. A mongrel at the lamppost lifts its leg, sniffs eagerly at the pavement and nimbly runs away. But what has all this to do with Liesbeth? Between her and the familiar everyday life yawns an unbridgeable abyss.

Mother is singing again. Liesbeth knows the song. It reaches her through the closed doors, disembodied, as if from far away. It is no longer as if Mother were singing; no, the notes speak for themselves, clear and melting and full of joy. They touch her wounds with tender, soothing fingers, the heavy stone in her heart melts and dissolves in a hot stream of tears.

The Young Girl

Les jeunes filles en fleur—*the staple of the bourgeois novel! And how many of these young girls are no more than plaster figurines, seen from the hero's perspective, designed by their creator to fall in love, marry in a positive waterfall of sweetness, then decline into the anonymity of motherhood!*

But there are the others—George Eliot's Maggie Tulliver, Colette's Vinca (The Ripening Seed), *Tolstoy's Natasha, Jane Austen's witty misses, and Zhenia, the protagonist of Pasternak's* Childhood of Luvers, *an astonishing artistic projection of the interior of a young girl's life.*

I have chosen two quite different selections. Olivia was written by Dorothy Bussy, a sister of Lytton Strachey, wife of the French painter Simon Bussy, and translator of Gide, and was published pseudonymously in 1949. It bears all the marks of an autobiographical experience preserved through the years and told in a brief novel of classic French clarity and proportion. Olivia is a young English girl sent to a school outside Paris run by Mademoiselle Cara and Mademoiselle Julie, where she is introduced to poetry, theater, music, and the semi-innocent love which blooms between teacher and pupil in such an enclosed female world. The school that Dorothy Bussy attended was the same school Eleanor Roosevelt went to which did so much for her, a school that was actually in England but was headed by a Frenchwoman who was the model for Mademoiselle Julie, one Mademoiselle Souvestre. Olivia's love for Mademoiselle Julie is an affair of the heart, not the flesh, for Julie, already tied to the jealous and hysterical Cara, then to the devoted Italian Signorina Baietto, shrinks from corrupting the child sent into her care. This concentrated moral tale breathes the essence of youthful sensuality turned aside from its natural expression and forced to flower in the overheated atmosphere of a gynaeceum (*see also the German film* Mädchen in Uniform).

Little Women, *that child of New England high thinking and simple living, is a world away from the European sophistication of Olivia. Who can measure the influence of Louisa May Alcott's novel on all of us who read it? Written for young girls, it remains a book of irresistible appeal when we pick it up twenty years after and plunge at once, with this opening chapter, into the family life of the Marches. Here are the four sisters, each embodying a type of femininity—calm, ladylike Meg; gifted, rebellious Jo; saintly, self-effacing Beth; and flirtatious Amy with her desire for lovely clothes and good social position. These full-scale portraits are the heart of the book, especially Jo, one of the first independent women in fiction and one of the most successfully realized. If we compare her, for example, with Anna, the heroine of* Doris Lessing's The Golden Notebook, *we see that, paradoxically, Jo is able to be independent of men in a way that Anna is not, for the restrictive sexual mores of the nineteenth century did not allow sexual freedom to substitute for a more profound inner freedom—which Jo does attain.*

Alcott lived during a time when women were victims of extreme social repression, when a respectable girl was supposed to stay at home until she married, and then stay in her husband's home. She might be a teacher or a governess, no more. Alcott, who never married, tried both jobs, tried everything possible, in her efforts to earn a living for herself and her family—headed by the impecunious transcendentalist Bronson Alcott—before she was able to make money as a writer. Over and over again, her books cry out for more independence for women and for an end to the irrational customs that confined them to such a restricted sphere.

OLIVIA (DOROTHY BUSSY)

Olivia

I was rather more than sixteen when my mother decided to take me away from Miss Stock's and send me for my "finishing" to a school in France. There was one already chosen to hand, kept by two French ladies whom my mother had met several years earlier when she was staying in a hotel in Italy, and who had remained her friends ever since.

Mademoiselle Julie T—— and Mademoiselle Cara M—— were dim figures flitting occasionally through my childhood, barely distinguishable from each other, but invested in a kind of romance from the fact of their foreign nationality. They sometimes came to stay with us a little in the holidays. They nearly always sent me a child's French book on New Year's Day. Starting with *Les Malheurs de Sophie,* we progressed gradually through several volumes of Erckmann-Chatrian up to *La Petite Fadette* and *François le Champi,* with one lurid and delightful interruption to dullness in the shape of a novel by Alphonse Daudet arranged for young people. Thanks to my mother and a French nursery maid, I knew French pretty well, that is I understood it when spoken and could read it fluently; but time was too precious to be wasted on French books, so that the only ones I read were my New Year presents, and those only as a matter of duty and politeness. At Miss Stock's, the French lessons, given by a deadly Mademoiselle, were a torture from which I took refuge as best I could in depths of agreeable abstraction, only coming to the surface for a moment when it was my turn to translate two or three lines of *L'Avare* or of whatever the classic might be we were spending that particular term in stumbling through.

The new school—Les Avons it was called—was situated in one of the loveliest parts of a great forest and within easy reach of Paris. It was delightful setting off for the first time abroad. I travelled with a party of other girls, some new and some old, under the conduct of the two ladies, "*ces dames*," as it was the fashion to call them. I can't remember much of the journey, except the excitement of it.

The school was a small one, consisting of not more than thirty girls, English, American and Belgian, and a staff of German, Italian, English and French mistresses, a music mistress, and so forth.

For the first time in my life I was given a delightful little bedroom entirely to myself, and I remember it was in that room that I first looked at myself in the glass—a proceeding for which the strictest privacy is necessary, and for which, to tell the truth, I had never felt much inclination. I was beginning the new life in very different circumstances from the old. Here, I was not going to be a pariah, a goat outside the pale of salvation, and looked at with suspicion and misgiving by the Wesleyan sheep gathered safely inside it. On the contrary, I was starting, I felt, with the sympathy of the authorities and the respect of my companions, the precious daughter of a highly revered friend; and if, thought I, there is such a friendship between the French ladies and my mother, it must be that they know her "views" and possibly share them.

"And who is that tiny thing like a brownie?" I asked next morning, as I watched a curious little figure tripping and bustling down the long broad passage.

"Oh, that's Signorina, the Italian mistress. She's on Mademoiselle Julie's side."

"And just think!" said someone else, "the German mistress is a *widow!*"

"Yes, and *she's* all for Mademoiselle Cara!"

Curious words. I didn't pay much attention to them, taken up as I was for the first few days by all the novelty around me, by the kind of disorder that reigned, by the chatter and laughter, by the foreign speech, by the absence of rules, by the extraordinary and delicious meals, by an atmosphere of gaiety and freedom which was like the breath of life to me.

It was the term that begins in spring and ends in summer, and I felt indeed as if I were coming to life with the rest of the world. The grip of a numbing winter was loosened, the frozen ground had thawed, the sun was shining, the air was soft, violets and primroses were pushing up their heads in the woods. The woods lay just on the other side of the road; when we went out for our walks, as soon as we had crossed it, we were allowed to break out of file and run about as we chose, pick flowers or play games. How beautiful the woods were! How different this was from those crocodile walks along the suburban, villa-lined roads round Stockhome, where we were not allowed to forget for a single moment that we were young ladies, but must walk in step and never fall out and not talk much, though talking was the only way of amusing oneself, for there was nothing about us that we cared to look at.

On that first morning walk, my companion was a lively, pretty girl called Mimi; she took with her on a lead a big St. Bernard dog who belonged to the school and whom she had special charge of. As soon as we got into the woods, she set him free, and the great creature rushed and bounded and tried to knock us down, and we laughed and shouted and were happy.

But though I enjoyed my walk, I wasn't sorry to go in. The first week at a new school is a busy one; curriculum to be talked over, time-tables to be arranged, names and faces to be learnt. Though a new girl, I at once took my place among the elder pupils. I knew French better than a great many of them; I was to attend the visiting professors' lectures and Mlle. Julie's literature lessons. (Mlle. Cara, I discovered, gave no lessons.) I was to begin Italian and go on with German and Latin; I was to be allowed to give up mathematics.

So far, Mlle. Julie and Mlle. Cara remained, as far as I was concerned, on their Olympian heights. I had very little to do with them and only distinguished one from the other by saying to myself that Mlle. Julie was the more lively and Mlle. Cara the kinder. One evening, my friend Mimi, the girl with the dog, said to me: "Mademoiselle Julie has gone to Paris and Mademoiselle Cara wants us to go and have coffee with her in her *cabinet de travail.* Go up now. I've something I must do, but I'll be there in a moment."

I went upstairs, quaking a little, for I remembered the terrifying solemnity of my visits to Miss Stock's private sitting room. But this, I thought, will probably be different. I hoped so.

Mlle. Cara's *cabinet de travail* was on the first floor, almost next door to my own bedroom and just opposite the "ladies'" apartment on the other side of the passage. I knocked at the door and was told to come in. Mlle. Cara was lying on a sofa, looking very pretty and invalidish,

I thought. Frau Riesener was bending over her, arranging a shawl over her feet. As I came in, I heard Mlle. Cara say: "No, no. No one cares how ill I am." Then she turned to me with a smile:

"Ah! There's Olivia. Come in, dear child. Sit down beside me and tell me what news you have from your dear Mamma."

Her voice was low, sweet and caressing, her manner all gentleness, all sympathy. She and Mlle. Julie, having known me from my childhood, always said *"tu"* to me. I liked it. There was something, I thought, very lovely in this habit of the French language which gives it an added grace, tenderness, *nuance,* sadly lacking in English, with its single use of "you."

Frau Riesener left the room almost at once, and when a minute or two later Mimi appeared, we were soon employed in half a dozen little ways. One of us had to fetch the eau-de-Cologne, the other soak a handkerchief and help the sufferer put it on her forehead to relieve the migraine; one had to fan her for a little, the other tuck up her shawl, which had slipped. But she was so grateful for all these little services that we enjoyed doing them and felt busy and happy. Then we had to serve the coffee and look in a cupboard for the box of chocolates; then Mimi was told to show me the album of school photographs. It was the most recent ones that I enjoyed looking at most, for among the many faces of old girls there were some of girls I could recognize as being still here. But it was an old girl's face that attracted me most. It stood out among the others, not for its beauty, for it was almost plain, but for its expression. I had never seen a face, I thought, so frank, so candid, so glad and so intelligent. But I couldn't analyse what charmed me so.

"Who is that?" I asked.

"Oh, Laura. Laura——" answered Mimi, and she said the name of a celebrated English statesman. "Yes, his daughter; she left last term."

After that, as the pages turned, it was her face I looked for in the groups, and exclaimed with pleasure as I found it:

"Laura! There's Laura!"

"Do you admire her?" asked Mlle. Cara. "For my part, I think she's downright ugly. No elegance. No grace. Always so dowdily dressed. But of course, she has inherited brains."

Mlle. Cara herself figured in all the photographs, graceful enough and languid, with a group of the smallest girls sitting at her feet.

"And Mademoiselle Julie?" I asked. "Why is she never there?"

"Oh, she hates being photographed. It's a mania."

And so the evening came to an end. It had been unlike any experience I had ever had of school and slid away very pleasantly, but —but—had I been altogether at my ease, hadn't I left Mlle. Cara's *cabinet de travail* with a curious little sensation of discomfort?

As we walked away down the long passage together, Mimi put her arm in mine.

"Mademoiselle Cara didn't like Laura," she said. "She was Mademoiselle Julie's favourite."

I had been at Les Avons about a week when, one evening after dinner, it was announced that Mlle. Julie was going to read to us.

Signorina came running up to me with sparkling eyes. She was almost as young as I was and I never looked upon her as a governess or a superior.

"Oh, Olivia mia, *chè piacere!* You'll like it. I know you will."

We collected in the big music room, dressed in our evening frocks, with or without needlework, as we preferred. I was surprised and relievéd to find there was no compulsion. After we had taken our seats, little Signorina flitted in and out among us, visiting those who were sewing and giving them advice, help, admiration, or scorn. I came in for the latter.

"Not like sewing!" she cried. "Great lazy one! Come and look at mine."

She took me up to a little stool which was placed close behind the tall straight armchair, evidently reserved for Mlle. Julie, and showed me her own piece of embroidery, so delicate, so filmy, so dainty, made of such exquisite lawn and adorned with such tiny stitches, that I exclaimed:

"Oh, but I'm not a fairy!"

As we were laughing together, Mlle. Julie came in; she gave Signorina a glance as she passed.

"Little vanity!" she said, and went on to her chair.

Signorina turned scarlet, took up her work with a dejected air, and was just going to sit down on her stool when Frau Riesener came in.

"Mademoiselle Cara wants you to make her tisane, Mademoiselle Baietto," she said. "You're the only person who makes it properly."

"Oh," said Signorina, "but I asked her before dinner whether she would be wanting it, and she said she wouldn't."

"Well," said Frau Riesener, "she wants it now."

Signorina cast an appealing glance at Mlle. Julie, who looked at her gravely and said:

"Go, my child."

Then, as Signorina went reluctantly out, Mlle. Julie took up her book and began turning over the pages. In the meantime, I had slipped back to my own seat at the other end of the room.

"I am going to read you Racine's *Andromaque*," said Mlle. Julie, "but before I begin, I'll ask you a few questions. Has anybody here ever heard of Andromaque?"

Apparently nobody ever had. At any rate there was a silence.

"Come, come," she said, "you can't all be as badly educated as that."

After another pause, I plucked up my courage and piped out:

"Hector's wife."

"Yes. And who was Orestes' father?"

I answered this too to her satisfaction. (Hadn't I browsed upon Pope's Homer since the age of twelve and eked it out with reading innumerable tales of Greek mythology?)

She went on with her questions and I answered them all until it came to Hermione.

"And Hermione?" she asked.

"I have never heard of Hermione."

"Ah!" she said. "Well, tonight you shall hear of her, and I hope never forget her. But as you've answered so well, come here and sit beside me."

She beckoned me up and made me take poor little Signorina's stool close to her elbow. Then, after lecturing us for a minute on the importance of mythology, she rapidly explained the situation at Pyrrhus's court, took up the book and began:

Oui, puisque je retrouve un ami si fidèle . . .

I have often wondered what share Racine had in lighting the flame that began to burn in my heart that night, or what share proximity. If she hadn't read just that play or if she hadn't called me up by chance to sit so near her, in such immediate contact, would the inflammable stuff which I carried so unsuspectingly within me have remained perhaps outside the radius of the kindling spark and never caught fire at all? But probably not; sooner or later, it was bound to happen.

There was a table in front of her with a lamp on it which cast its light on her book and her face. I, sitting beside and below her, saw her illuminated and almost in profile. I looked at her for the first time as I listened. I don't know which I did more thirstily—looked or listened. It suddenly dawned upon me that this was beauty—great beauty—a thing I had read of and heard of without understanding, a thing I had passed by perhaps a hundred times with careless, unseeing eyes. Pretty girls I had seen, lovely girls, no doubt, but I had never paid much conscious attention to their looks, never been particularly interested in them. But this was something different. No, it was not different. It was merely being awakened to something for the first time—physical beauty. I was never blind to it again.

Who can describe a face? Who can forbear trying to? But such descriptions resolve themselves into an inventory of items. As item: a rather broad face, a low forehead, dark hair with a thread or two of grey in it, parted in the middle, gently waving on the temples and gathered up into a bunch of curls at the back of the head. A curious kind of hairdressing which I have never seen except in pictures or statues. The features were regular, cleanly cut and delicately formed, nose, lips, and chin fine and firm. The eyes were grey, sometimes clear and translucid, sometimes dark, impenetrable, burning. It was thanks to Racine that night that I saw a little of what they could express.

What a strange relationship exists between the reader and his listener. What an extraordinary breaking down of barriers. The listener is suddenly given the freedom of a city at whose gates he would never have dreamt of knocking. He may enter forbidden precincts. He may communicate at the most sacred altars with a soul he has never dared, never will dare approach, watch without fear or shame a spirit that has dropped its arms, its veils, its prudences, its reserves. He who is not beloved may gaze and hearken and learn at last what nothing else will ever reveal to him and what he longs to know even at the cost of life itself—how the beloved face is moved by passion, how scorn sits upon those features, and anger and love. How the beloved's voice softens and trembles into tenderness or breaks in the anguish of jealousy and grief. . . . Oh, but it is too soon to say all this. All these are reflections of a later date.

I have heard many readers read Racine, and famous men among them, but I have not heard any who read him as well as Mlle. Julie. She read simply and rapidly, without any of the actor's arts and affectations, with no swelling voice, with no gestures beyond the occasional lifting of her hand, in which she held a long ivory paper cutter. But the gravity of her bearing and her voice transported me at once into the courts of princes and the presence of great emotions:

> *Avant que tous les Grecs vous parlent par ma voix,*
> *Souffrez que j'ose ici me flatter de leur choix,*
> *Et qu'à vos yeux, Seigneur, je montre quelque joie*
> *De voir le fils d'Achille et le vainqueur de Troye . . .*

The sonorous vowels, the majestic periods, the tremendous names sweep on; one is borne upon a tide of music and greatness; one follows breathlessly the evolutions, the shiftings, the advances and retreats of the doomed quartet as they tread their measured way to death and madness, through all the vicissitudes of irresolution, passion and jealousy, leaving at the end a child's soul shaken and exhausted, the first great rent made in the veil that hides the emotions of men and women from the eyes of innocence.

Did I understand the play at that first reading? Oh, certainly not. Haven't I put the gathered experience of years into my recollection of it? No doubt. What is certain is that it gave me my first conception of tragedy, of the terror and complication and pity of human lives. Strange that for an English child that revelation should have come through Racine instead of through Shakespeare. But it did.

I went to bed that night in a kind of daze, slept as if I had been drugged and in the morning awoke to a new world—a world of excitement—a world in which everything was fierce and piercing, everything charged with strange emotions, clothed with extraordinary mysteries, and in which I myself seemed to exist only as an inner core of palpitating fire.

The walk that morning, the beauty of the forest, the sky, the deliciousness of the air, the delight of running—for the first time I enjoyed these things consciously.

"I understand," I cried to myself, "I understand at last. Life, life, life, this is life, full to overflowing with every ecstasy and every agony. It is mine, mine to hug, to exhaust, to drain."

And lessons! I went to them with a renovated ardour. Oh yes, I had been a fairly intelligent pupil; I had enjoyed learning and working in a kind of humdrum way. This was something quite different—something I had never known. Every page of the Latin grammar seemed to hold some passionate secret which must be mine or I should die. Words! How astonishing they were. The simplest bore with it such an aura of music and romance as wafted me into fairyland. Geography! Oh, to sit poring and wondering over an atlas. Here were pagodas. There the Nile. Jungles. Deserts. Coral islands in the Pacific ringed round with lagoons. The eternal snows of the Himalayas! Aurora Borealis flaming at the pole! Worlds upon worlds of magic revealed! Why had I never known of them before? History! Those men! Those heroes! How they looked, how they smiled as they were going to the block or the stake! And what had they died for? Faith, liberty, truth, humanity. What did those words really represent? I mustn't rest till I found out. And the peoples! The poor sheeplike peoples! Those too must be thought of. Not yet. I dare not yet. There will be time enough for that later. I am not strong enough yet to look really at all those dreadful meaningless pains. I must put that at the back of my mind. Now, now, I must grow strong. I must feed on beauty and rapture in order to grow strong.

And first of all that face. There was that to look at. A long way off, at the end of a table. Passing one on the stairs, coming suddenly out of a door. Talking to other people. Listening to other people. And sometimes, rarely, reading aloud. Had I then never looked at a face before? Why should the mere sight of it make my heart stand still? What was there so extraordinarily fascinating in watching it? Was it more satisfying when it was motionless, when one could imprint the line of the profile on one's memory, so fine and grave and austere, the delicate curl of the lip, the almost imperceptible and indescribably touching faint hollow of the cheek, the fall of the lashes on the pale skin, the curve of the dark hair on the brow? Or was it when expressions flowed over it so swiftly that one's eyes and one's heart were never quick enough to register them? Laughter was never long absent from it, spreading from the slight quiver of a smile to a ripple, to a tempest of gaiety, passing like a flash of lightning, a flood of colour, transforming, vivifying every feature. So I watched from afar. At meals especially, where I sat some way off but on the opposite side of the table.

There were three tables in the big dining room; the two heads, at the centre one, sat opposite to each other, as the foreign fashion is, in the middle of each long side. When there were guests or visiting professors, they sat on each side of the ladies. Special dishes were generally served for these honoured ones, and if any remained over, the servant was told to hand it round to the young ladies. Once when this occurred, Mlle. Julie cross-examined the girls who had been served in this way:

"Did you like that dish? Honestly now. As much as your English roast beef? No—? Yes—? You don't know? Ah, these English! They have no taste. And you, Olivia, what did you think of it?"

My answer, "Delicious!" was so fervent that she laughed:

"Ha! Have we got a gourmet at last? But appreciation isn't all. There must be discrimination too. Was there anything in the dish that you think might be criticised? Anything that might have improved it?"

"I think—" I murmured.

"Yes, out with it."

"There was perhaps a thought too much lemon in it."

"Bravo!" she cried. "You deserve encouragement. You shall be promoted."

And at the next meal, after an anxious search, I found my napkin ring had been placed next to Mlle. Julie's own. And it was there, at her right hand, that I sat till the end of my time at Les Avons, unless a visitor or a professor sometimes separated us. And now she almost always helped me herself to one or other of the special dishes, calling me "Mademoiselle Gourmet," asking me my opinion, laughing at my enjoyment, teasing me for being still too "English," because I wouldn't drink wine. "But perhaps," she said, "our *vin ordinaire* isn't good enough for you?" And perhaps, indeed, that was it.

But there was no need of wine to intoxicate me. Everything in her proximity was intoxicating. And I was now, for the first time, within range of her talk. Mlle. Julie's talk, I discovered later, was celebrated, and not only amongst us schoolgirls, but amongst famous men, whose names we whispered.

I had no doubt been accustomed, or ought to have been accustomed, to good talk at home. But at home one was inattentive. There were all the other children who somehow interfered. It was on their level, in their turmoil, that one lived. They were too distracting to allow of one's taking any interest in one's elders and their conversation. When one did listen to it, it was mostly political, or else took the form of argument. My mother and my aunt, who was often in the house, had interminable and heated discussions, in which my mother was invariably in the right and my aunt beyond belief inconsequent and passionate. We found them tedious and sometimes nerve-racking. My father, a man, in our eyes, of infinite wisdom and humour, did not talk much; he was fond of explaining scientific or mathematical problems to us, or, occasionally, of inventing and making us take a share in some fantastic piece of tomfoolery. He would let fall from time to time a grim and gnomic apothegm, which we treasured as a household word, and would often calm a heated discussion by an apparently irrelevant absurdity. As for the people who came to the house, many of whom were highly distinguished, we admired them without listening to them. Their world seemed hardly to impinge upon ours.

How different it was here! Mlle. Julie was witty. Her brilliant speech darted here and there with the agility and grace of a hummingbird. Sharp and pointed, it would sometimes transfix a victim cruelly. No one was safe, and if one laughed with her, one was liable the next minute to be pierced oneself with a shaft of irony. But she tossed her epigrams about with such evident enjoyment, that if one had the smallest sense of fun, one enjoyed them too, and it was from her that I, for one, learnt to realize the exquisite adaptation of the French tongue to the French wit. But her talk was not all epigrams. One felt it informed by that infectious ardour, that enlivening zest, which were the secret of her success as a schoolmistress. There was nothing into which she could not infuse them. Every subject, however dull it had seemed in the hands of others, became animated in hers. With the traditional culture of a French Protestant family, having contacts with eminent men and women in many countries, she had too a spontaneous and open mind, capable of points of view, fond of the stimulus of paradox. The dullest of her girls was stirred into some sort of life in her presence; to the intelligent, she communicated a Promethean fire which warmed and coloured their whole lives. To sit at table at her right hand was an education itself.

LOUISA MAY ALCOTT
(1 8 3 2 – 1 8 8 8)

Little Women

"Christmas won't be Christmas without any presents," grumbled Jo, lying on the rug.

"It's so dreadful to be poor!" sighed Meg, looking down at her old dress.

"I don't think it's fair for some girls to have plenty of pretty things, and other girls nothing at all," added little Amy, with an injured sniff.

"We've got father and mother and each other," said Beth contentedly, from her corner.

The four young faces on which the firelight shone brightened at the cheerful words, but darkened again as Jo said sadly,—"We haven't got father, and shall not have him for a long time." She didn't say "perhaps never," but each silently added it, thinking of father far away, where the fighting was.

Nobody spoke for a minute; then Meg said in an altered tone,—

"You know the reason mother proposed not having any presents this Christmas was because it is going to be a hard winter for everyone; and she thinks we ought not to spend money for pleasure, when our men are suffering so in the army. We can't do much, but we can make our little sacrifices, and ought to do it gladly. But I am afraid I don't;" and Meg shook her head, as she thought regretfully of all the pretty things she wanted.

"But I don't think the little we should spend would do any good. We've each got a dollar, and the army wouldn't be much helped by our giving that. I agree not to expect anything from mother or you, but I do want to buy Undine and Sintram for myself; I've wanted it *so* long," said Jo, who was a bookworm.

"I have planned to spend mine in new music," said Beth, with a little sigh, which no one heard but the hearth-brush and kettle-holder.

"I shall get a nice box of Faber's drawing-pencils; I really need them," said Amy decidedly.

"Mother didn't say anything about our money, and she won't wish us to give up everything. Let's each buy what we want, and have a little fun; I'm sure we work hard enough to earn it," cried Jo, examining the heels of her shoes in a gentlemanly manner.

"I know *I* do,—teaching those tiresome children nearly all day, when I'm longing to enjoy myself at home," began Meg, in the complaining tone again.

"You don't have half such a hard time as I do," said Jo. "How would you like to be shut up for hours with a nervous, fussy old lady, who keeps you trotting, is never satisfied, and worries you till you're ready to fly out of the window or cry?"

"It's naughty to fret; but I do think washing dishes and keeping things tidy is the worst work in the world. It makes me cross; and my hands get so stiff, I can't practise well at all"; and Beth looked at her rough hands with a sigh that any one could hear that time.

"I don't believe any of you suffer as I do," cried Amy; "for you don't have to go to school with impertinent girls, who plague you if you don't know your lessons, and laugh at your dresses, and label your father if he isn't rich, and insult you when your nose isn't nice."

"If you mean *libel,* I'd say so, and not talk about *labels,* as if papa was a pickle-bottle," advised Jo, laughing.

"I know what I mean, and you needn't be *statirical* about it. It's proper to use good words, and improve your *vocabilary,*" returned Amy, with dignity.

"Don't peck at one another, children. Don't you wish we had the money papa lost when we were little, Jo? Dear me! how happy and good we'd be, if we had no worries!" said Meg, who could remember better times.

"You said, the other day, you thought we were a deal happier than the King children, for they were fighting and fretting all the time, in spite of their money."

"So I did, Beth. Well, I think we are; for, though we do have to work, we make fun for ourselves, and are a pretty jolly set, as Jo would say."

"Jo does use such slang words!" observed Amy, with a reproving look at the long figure stretched on the rug. Jo immediately sat up, put her hands in her pockets, and began to whistle.

"Don't, Jo; it's so boyish!"

"That's why I do it."

"I detest rude, unlady-like girls!"

"I hate affected, niminy-piminy chits!"

" 'Birds in their little nests agree,' " sang Beth, the peacemaker, with such a funny face that both sharp voices softened to a laugh, and the "pecking" ended for that time.

"Really, girls, you are both to be blamed," said Meg, beginning to lecture in her elder-sisterly fashion. "You are old enough to leave off boyish tricks, and to behave better, Josephine. It didn't matter so much when you were a little girl; but now you are so tall, and turn up your hair, you should remember that you are a young lady."

"I'm not! and if turning up my hair makes me one, I'll wear it in two tails till I'm twenty," cried Jo, pulling off her net, and shaking down a chestnut mane. "I hate to think I've got to grow up, and be Miss March, and wear long gowns, and look as prim as a China-aster! It's bad enough to be a girl, anyway, when I like boys' games and work and manners! I can't get over my disappointment in not being a boy; and it's worse than ever now, for I'm dying to go and fight with papa, and I can only stay at home and knit, like a poky old woman!" And Jo shook the blue army-sock till the needles rattled like castanets, and her ball bounded across the room.

"Poor Jo! It's too bad, but it can't be helped; so you must try to be contented with making your name boyish, and playing brother to us girls," said Beth, stroking the rough head at her knee with a hand that all the dish-washing and dusting in the world could not make ungentle, in its touch.

"As for you, Amy," continued Meg, "you are altogether too particular and prim. Your airs are funny now; but you'll grow up an affected little goose, if you don't take care. I like your nice manners and refined ways of speaking, when you don't try to be elegant; but your absurd words are as bad as Jo's slang."

"If Jo is a tom-boy and Amy a goose, what am I, please?" asked Beth, ready to share the lecture.

"You're a dear, and nothing else," answered Meg warmly; and no one contradicted her, for the "Mouse" was the pet of the family.

As young readers like to know "how people look," we will take this moment to give them a little sketch of the four sisters, who sat knitting away in the twilight, while the December snow fell quietly without, and the fire crackled cheerfully within. It was a comfortable old room, though the carpet was faded and the furniture very plain; for a good picture or two hung on the walls, books filled the recesses, chrysanthemums and Christmas roses bloomed in the windows, and a pleasant atmosphere of home-peace pervaded it.

Margaret, the eldest of the four, was sixteen, and very pretty, being plump and fair, with large eyes, plenty of soft, brown hair, a sweet mouth, and white hands, of which she was rather vain. Fifteen-year-old Jo was very tall, thin, and brown, and reminded one of a colt; for she never seemed to know what to do with her long limbs, which were very much in her way. She had a decided mouth, a comical nose, and sharp, gray eyes, which appeared to see everything, and were by turns fierce, funny, or thoughtful. Her long, thick hair was her one beauty; but it was usually bundled into a net, to be out of her way. Round shoulders had Jo, big hands and feet, a fly-away look to her clothes, and the uncomfortable appearance of a girl who was rapidly shooting up into a woman, and didn't like it. Elizabeth—or Beth, as everyone called her—was a rosy, smooth-haired, bright-eyed girl of thirteen, with a shy manner, a timid voice, and a peaceful expression, which was seldom disturbed. Her father called her "Little Tranquillity," and the name suited her excellently; for she seemed to live in a happy world of her own, only venturing out to meet the few whom she trusted and loved. Amy, though the youngest, was a most important person,—in her own opinion at least. A regular snow-maiden, with blue eyes, and yellow hair, curling on her shoulders, pale and slender, and always carrying herself like a young lady mindful of her manners. What the characters of the four sisters were we will leave to be found out.

The clock struck six; and, having swept up the hearth, Beth put a pair of slippers down to warm. Somehow the sight of the old shoes had a good effect upon the girls; for mother was coming, and everyone brightened to welcome her. Meg stopped lecturing, and lighted the lamp, Amy got out of the easy-chair without being asked, and Jo forgot how tired she was as she sat up to hold the slippers nearer to the blaze.

"They are quite worn out; Marmee must have a new pair."

"I thought I'd get her some with my dollar," said Beth.

"No, I shall!" cried Amy.

"I'm the oldest," began Meg, but Jo cut in with a decided—

"I'm the man of the family now papa is away, and *I* shall provide the slippers, for he told me to take special care of mother while he was gone." .

"I'll tell you what we'll do," said Beth; "let's each get her something for Christmas, and not get anything for ourselves."

"That's like you, dear! What will we get?" exclaimed Jo.

Everyone thought soberly for a minute; then Meg announced, as if the idea was suggested by the sight of her own pretty hands, "I shall give her a nice pair of gloves."

"Army shoes, best to be had," cried Jo.

"Some handkerchiefs, all hemmed," said Beth.

"I'll get a little bottle of cologne; she likes it, and it won't cost much, so I'll have some left to buy my pencils," added Amy.

"How will we give the things?" asked Meg.

"Put them on the table, and bring her in and see her open the bundles. Don't you remember how we used to do on our birthdays?" answered Jo.

"I used to be *so* frightened when it was my turn to sit in the big chair with the crown on, and see you all come marching round to give the presents, with a kiss. I liked the things and the kisses, but it was dreadful to have you sit looking at me while I opened the bundles," said Beth, who was toasting her face and the bread for tea, at the same time.

"Let Marmee think we are getting things for ourselves, and then surprise her. We must go shopping to-morrow afternoon, Meg; there is so much to do about the play for Christmas night," said Jo, marching up and down, with her hands behind her back and her nose in the air.

"I don't mean to act any more after this time; I'm getting too old for such things," observed Meg, who was as much a child as ever about "dressing-up" frolics.

"You won't stop, I know, as long as you can trail round in a white gown with your hair down, and wear gold-paper jewelry. You are the best actress we've got, and there'll be an end of everything if you quit the boards," said Jo. "We ought to rehearse to-night. Come here,

Amy, and do the fainting scene, for you are as stiff as a poker in that."

"I can't help it; I never saw anyone faint, and I don't choose to make myself all black and blue, tumbling flat as you do. If I can go down easily, I'll drop; if I can't I shall fall into a chair and be graceful; I don't care if Hugo does come at me with a pistol," returned Amy, who was not gifted with dramatic power, but was chosen because she was small enough to be borne out shrieking by the villain of the piece.

"Do it this way: clasp your hands so, and stagger across the room, crying frantically, 'Roderigo! save me! save me!'" and away went Jo with a melodramatic scream which was truly thrilling.

Amy followed, but she poked her hands out stiffly before her, and jerked herself along as if she went by machinery; and her "Ow!" was more suggestive of pins being run into her than of fear and anguish. Jo gave a despairing groan, and Meg laughed outright, while Beth let her bread burn as she watched the fun, with interest.

"It's no use! Do the best you can when the time comes and if the audience laugh, don't blame me. Come on, Meg."

Then things went smoothly, for Don Pedro defied the world in a speech of two pages without a single break; Hagar, the witch, chanted an awful incantation over her kettleful of simmering toads, with weird effect; Roderigo rent his chains asunder manfully, and Hugo died in agonies of remorse and arsenic, with a wild "Ha! ha!"

"It's the best we've had yet," said Meg, as the dead villain sat up and rubbed his elbows.

"I don't see how you can write and act such splendid things, Jo. You're a regular Shakespeare!" exclaimed Beth, who firmly believed that her sisters were gifted with wonderful genius in all things.

"Not quite," replied Jo modestly. "I do think 'The Witch's Curse, an Operatic Tragedy,' is rather a nice thing; but I'd like to try Macbeth, if we only had a trap-door for Banquo. I always wanted to do the killing part. 'Is that a dagger that I see before me?'" muttered Jo, rolling her eyes and clutching at the air, as she had seen a famous tragedian do.

"No, it's the toasting fork, with mother's shoe on it instead of the bread. Beth's stage-struck!" cried Meg, and the rehearsal ended in a general burst of laughter.

"Glad to find you so merry, my girls," said a cheery voice at the door, and actors and audience turned to welcome a tall, motherly lady, with a "can-I-help-you" look about her which was truly delightful. She was not elegantly dressed, but a noble-looking woman, and the girls thought the gray cloak and unfashionable bonnet covered the most splendid mother in the world.

"Well, dearies, how have you got on to-day? There was so much to do, getting the boxes ready to go to-morrow, that I didn't come home to dinner. Has anyone called, Beth? How is your cold, Meg? Jo, you look tired to death. Come and kiss me, baby."

While making these maternal inquiries Mrs. March got her wet things off, her warm slippers on, and sitting down in the easy-chair, drew Amy to her lap, preparing to enjoy the happiest hour of her busy day. The girls flew about, trying to make things comfortable, each in her own way. Meg arranged the tea-table; Jo brought wood and set chairs, dropping, overturning, and clattering everything she touched; Beth trotted to and fro between parlor and kitchen, quiet and busy; while Amy gave directions to everyone, as she sat with her hands folded.

As they gathered about the table, Mrs. March said, with a particularly happy face, "I've got a treat for you after supper."

A quick, bright smile went round like a streak of sunshine. Beth clapped her hands, regardless of the biscuit she held, and Jo tossed up her napkin, crying, "A letter! a letter! Three cheers for father!"

"Yes, a nice long letter. He is well, and thinks he shall get through the cold season better than we feared. He sends all sorts of loving wishes for Christmas, and an especial message to you girls," said Mrs. March, patting her pocket as if she had got a treasure there.

"Hurry and get done! Don't stop to quirk your little finger, and simper over your plate, Amy," cried Jo, choking in her tea, and dropping her bread, butter side down, on the carpet, in her haste to get at the treat.

Beth ate no more, but crept away, to sit in her shadowy corner and brood over the delight to come, till the others were ready.

"I think it was so splendid in father to go as a chaplain when he was too old to be drafted, and not strong enough for a soldier," said Meg warmly.

"Don't I wish I could go as a drummer, a *vivan*—what's its name?

or a nurse, so I could be near him and help him," exclaimed Jo, with a groan.

"It must be very disagreeable to sleep in a tent, and eat all sorts of bad-tasting things, and drink out of a tin mug," sighed Amy.

"When will he come home, Marmee?" asked Beth, with a little quiver in her voice.

"Not for many months, dear, unless he is sick. He will stay and do his work faithfully as long as he can, and we won't ask for him back a minute sooner than he can be spared. Now come and hear the letter."

They all drew to the fire, mother in the big chair with Beth at her feet, Meg and Amy perched on either arm of the chair, and Jo leaning on the back, where no one would see any sign of emotion if the letter should happen to be touching. Very few letters were written in those hard times that were not touching, especially those which fathers sent home. In this one little was said of the hardships endured, the dangers faced, or the homesickness conquered; it was a cheerful, hopeful letter, full of lively descriptions of camp life, marches, and military news; and only at the end did the writer's heart overflow with fatherly love and longing for the little girls at home.

"Give them all my dear love and a kiss. Tell them I think of them by day, pray for them by night, and find my best comfort in their affection at all times. A year seems very long to wait before I see them, but remind them that while we wait we may all work, so that these hard days need not be wasted. I know they will remember all I said to them, that they will be loving children to you, will do their duty faithfully, fight their bosom enemies bravely, and conquer themselves so beautifully, that when I come back to them I may be fonder and prouder than ever of my little women."

Everybody sniffed when they came to that part; Jo wasn't ashamed of the great tear that dropped off the end of her nose, and Amy never minded the rumpling of her curls as she hid her face on her mother's shoulder and sobbed out, "I *am* a selfish girl! but I'll truly try to be better, so he mayn't be disappointed in me by and by."

"We all will!" cried Meg. "I think too much of my looks, and hate to work, but won't any more, if I can help it."

"I'll try and be what he loves to call me, 'a little woman,' and not be rough and wild; but do my duty here instead of wanting to be somewhere else," said Jo, thinking that keeping her temper at home was a much harder task than facing a rebel or two down South.

Beth said nothing, but wiped away her tears with the blue army-sock, and began to knit with all her might, losing no time in doing the duty that lay nearest her, while she resolved in her quiet little soul to be all that father hoped to find her when the year brought round the happy coming home.

Mrs. March broke the silence that followed Jo's words, by saying in her cheery voice, "Do you remember how you used to play Pilgrim's Progress when you were little things? Nothing delighted you more than to have me tie my piece-bags on your backs for burdens, give you hats and sticks and rolls of paper, and let you travel through the house from the cellar, which was the City of Destruction, up, up, to the house-top, where you had all the lovely things you could collect to make a Celestial City."

"What fun it was, especially going by the lions, fighting Apollyon, and passing through the Valley where the hobgoblins were!" said Jo.

"I liked the place where the bundles fell off and tumbled down-stairs," said Meg.

"My favorite part was when we came out on the flat roof where our flowers and arbors and pretty things were, and all stood and sung for joy up there in the sunshine," said Beth, smiling, as if that pleasant moment had come back to her.

"I don't remember much about it, except that I was afraid of the cellar and the dark entry, and always liked the cake and milk we had up at the top. If I wasn't too old for such things, I'd rather like to play it over again," said Amy, who began to talk of renouncing childish things at the mature age of twelve.

"We never are too old for this, my dear, because it is a play we are playing all the time in one way or another. Our burdens are here, our road is before us, and the longing for goodness and happiness is the guide that leads us through many troubles and mistakes to the peace which is a true Celestial City. Now, my little pilgrims, suppose you begin again, not in play, but in earnest, and see how far on you can get before father comes home."

"Really, mother? Where are our bundles?" asked Amy, who was a very literal young lady.

"Each of you told what your burden was just now, except Beth; I rather think she hasn't got any," said her mother.

"Yes, I have; mine is dishes and dusters, and envying girls with nice pianos, and being afraid of people."

Beth's bundle was such a funny one that everybody wanted to laugh; but nobody did, for it would have hurt her feelings very much.

"Let us do it," said Meg thoughtfully. "It is only another name for trying to be good, and the story may help us; for though we do want to be good, it's hard work, and we forget, and don't do our best."

"We were in the Slough of Despond to-night, and mother came and pulled us out as Help did in the book. We ought to have our roll of directions, like Christian. What shall we do about that?" asked Jo, delighted with the fancy which lent a little romance to the very dull task of doing her duty.

"Look under your pillows, Christmas morning, and you will find your guide-book," replied Mrs. March.

They talked over the new plan while old Hannah cleared the table; then out came the four little work-baskets, and the needles flew as the girls made sheets for Aunt March. It was uninteresting sewing, but to-night no one grumbled. They adopted Jo's plan of dividing the long seams into four parts, and calling the quarters Europe, Asia, Africa, and America, and in that way got on capitally, especially when they talked about the countries as they stitched their way through them.

At nine they stopped work, and sung, as usual, before they went to bed. No one but Beth could get much music out of the old piano; but she had a way of softly touching the yellow keys, and making a pleasant accompaniment to the simple songs they sung. Meg had a voice like a flute, and she and her mother led the little choir. Amy chirped like a cricket, and Jo wandered through the airs at her own sweet will, always coming out at the wrong place with a croak or a quaver that spoilt the most pensive tune. They had always done this from the time they could lisp "Crinkle, crinkle, 'ittle 'tar," and it had become a household custom, for the mother was a born singer. The first sound in the morning was her voice, as she went about the house singing like a lark; and the last sound at night was the same cheery sound, for the girls never grew too old for that familiar lullaby.

The Virgin

Clarissa *is the immediate successor of one of the first true novels in English,* Pamela, *also written by Samuel Richardson. A printer who had done well with a book giving examples of proper letter styles designed for use by the growing merchant class, Richardson went on to compose epistolary novels which were great successes in his day. While neither* Pamela *nor his other novel,* Charles Grandison, *can hold our attention today,* Clarissa *continues to fascinate. Introducing it to a class of college seniors, I discovered that they were caught up in it despite the strangeness of form and style, and eagerly turned its pages to find the answer to the single question on which Richardson hangs the story: "Did she or didn't she?"*

Is there another book which sticks so obsessively to physical virginity as the keystone of everything else in life worth having? By enlarging until it fills the entire foreground the Christian belief in the supreme value of virginity at the moment when it shades into the merchant concept of virginity as a marketable commodity, Richardson expressed an almost mythic vision of life, one which kept women barricaded within the triple barriers of their homes, their clothes, and their untouchable skins. A nervous sexuality pervades every page—reading it for the first time after reading twentieth-century novels, one sees its relationships with Proust or with the more brilliant and subtle eighteenth-century Les Liaisons dangereuses *of the French Choderlos de Laclos, where sexual corruption is anatomized with a scalpel honed to draw blood.*

English Puritanism forced much of Clarissa's sexuality underground, but it is not invisible for all of that. Clarissa protests her innocence with every letter; still, her actions are ambiguous, there is good reason

84

for her father's apprehensions. His sin is to see Clarissa not as a person but as property, ready to be matched to a suitable male to insure continued commercial prosperity for the Harlowe family.

It is possible to dip into this enormously long book anywhere and come up with the same combination of hysterical emotion and barely controlled sensuality. The letters which follow have been written by Clarissa to her dear friend Anna Howe after Clarissa has incurred the wrath of her father and brother by refusing to marry Mr. Solmes, their choice for her, an admirable, dull, ugly man, at least to Clarissa. Her father wishes her to marry because he fears her involvement with Lovelace, a handsome, dangerous man who desires Clarissa but balks at associating himself, an aristocrat, with the mercantile pretensions of the Harlowes. Clarissa vows herself to perpetual virginity and pleads obedience to all her father's wishes but this one—she will not marry Mr. Solmes. She weeps, faints, throws herself at his feet, refuses the pleas of her mother, and carries on at a pitch of exhausting hysteria. Finally she is confined to her room. Her father is not entirely simple, however, for Clarissa does manage to communicate with Miss Howe, as in these letters, and through her with Lovelace, who will soon elope with Clarissa to London.

The relationship between Clarissa and Lovelace is drawn in fine detail. Maddened by her icy resistance, Lovelace is determined to force Clarissa and bring her to acknowledge his manhood, his power, his flesh. He wants to dirty her and make her human. Although she puts herself in his power, Clarissa will accept him only on her terms, as a defanged suppliant for her favors. In the pages of Clarissa, then, is played out one of the archetypal dramas of the sexual relation. It is also the completest depiction of the myth of the Virgin; the largely insipid virgins of the nineteenth-century fiction are but fragments of Clarissa's far from pallid character.

SAMUEL RICHARDSON
(1689–1761)

Clarissa

Miss Clarissa Harlowe to Miss Howe

Sat., March 4. 12 o'clock.

As to what you mention of my sister's value for Mr. Lovelace, I am not very much surprised at it. She never tells the story of their parting, and of her refusal of him, but her colour rises, she looks with disdain upon me, and mingles anger with the airs she gives herself: anger as well as airs, demonstrating that she refused a man whom she thought worth accepting: where else is the reason either for anger or boast?

As to what you say of my giving up to my father's control the estate devised me, my motives at the time, as you acknowledge, were not blameable. You were indeed jealous of my brother's views *against me;* or rather of his predominant love of *himself;* but I did not think so hardly of my brother and sister as you always did.

And now for the *most* concerning part of your letter.

You think I must of necessity, as matters are circumstanced, be Solmes's wife. I will not be very rash, my dear, in protesting to the contrary: but I think it never can, and what is still more, never *ought* to be! I repeat, that I *ought* not: for surely, my dear, I should not give up to my brother's ambition the happiness of my future life. The less, surely, ought I to give into these grasping views of my brother, as I myself heartily despise the end aimed at; as I wish not either to change my state, or better my fortunes; and as I am fully persuaded, that happiness and riches are *two* things and very seldom meet together.

I am stopped. Hannah shall deposit this. She was ordered by my mother (who asked where I was) to tell me that she would come up and talk with me in my own closet. She is coming! Adieu, my dear.

Miss Clarissa Harlowe to Miss Howe

Sat. Afternoon.

The expected conference is over: but my difficulties are increased.

This, as my mother was pleased to tell me, being the last *persuasory* effort that is to be attempted.

I have made, said she, as she entered my room, a *short* as well as *early* dinner, on purpose to confer with you: and I do assure you that it will be the last conference I shall either be permitted or *inclined* to hold with you on the subject, if you should prove as refractory as it is imagined you will prove by some, who are of opinion that I have not the weight with you which my indulgence deserves. But I hope you will convince as well them as me of the contrary.

Your father both dines and sups at your uncle's, on purpose to give us this opportunity; and according to the report I shall make on his return (which I have promised shall be a very faithful one), he will take his measures with you.

I was offering to speak—hear, Clarissa, what I have to tell you, said she, before you speak, unless what you have to say will signify to me your compliance—say—*will* it? If it *will,* you may speak.

I was silent.

She looked with concern and anger upon me—no compliance, I find! —such a dutiful young creature hitherto! Will you not, *can* you not, speak as I would have you speak? Then (rejecting me as if it were with her hand) continue silent. *I,* no more than your *father,* will bear your *avowed* contradiction.

She paused, with a look of expectation, as if she waited for my consenting answer.

I was still silent, looking down; the tears in my eyes.

O thou determined girl! But say—speak out—are you resolved to stand in opposition to us all, in a point our hearts are set upon?

May I, madam, be permitted to expostulate?

To what purpose expostulate with *me,* Clarissa? Your *father* is determined. Have I not told you there is no receding; that the honour as well as the interest of the family is concerned? Be ingenuous: you *used* to be so, even occasionally against yourself: who at the long run *must* submit—*all* of us to *you;* or *you* to *all* of us? If you intend to yield at *last* if you find you cannot conquer, yield *now,* and with a grace—for yield you must, or be none of our child.

I wept. I knew not what to say; or rather how to express what I had to say.

Take notice, that there are flaws in your grandfather's will: not a

shilling of that estate will be yours if you do not yield. Your grandfather left it to you as a reward of your duty to *him* and to *us*. You will
justly forfeit it, if——

Permit me, good madam, to say that, if it were *unjustly* bequeathed
me, I ought not to wish to have it. But I hope Mr. Solmes will be
apprised of these flaws.

This is very pertly said, Clarissa: but reflect, that the forfeiture of
that estate through your opposition will be attended with the total
loss of your father's favour.

I must accommodate myself, madam. It becomes me to be thankful
for what I have had.

What perverseness! said my mother. But if you depend upon the
favour of either or both your uncles, vain will be that dependence:
they will give you up, I do assure you, if your father does, and absolutely renounce you.

I am sorry, madam, that I have had so little merit as to have made
no deeper impressions of favour for me in their hearts: but I will love
and honour them as long as I live.

All this, Clarissa, makes your prepossession in a certain man's favour the more evident. Indeed your brother and sister cannot go any
whither but they hear of these prepossessions.

It is a great grief to me, madam, to be made the subject of the public
talk: but I hope you will have the goodness to excuse me for observing
that the authors of my disgrace within doors, the talkers of my prepossession without, and the reporters of it from abroad, are originally
the same persons.

She severely chid me for this.

I received her rebukes in silence.

You are sullen, Clarissa: I see you are *sullen*. And she walked
about the room in anger. Then turning to me—you can *bear* the
imputation of sullenness, I see! I was afraid of telling you all I was en-
joined to tell you, in case you were to be unpersuadable: but I find that
I had a greater option of your delicacy, of your gentleness than I
needed to have—it cannot discompose so steady, so inflexible a young
creature, to be told, as I now tell you, that the settlements are actually
drawn; and that you will be called down in a very few days to hear
them read and to sign them: for it is impossible, if your heart be free,
that you can make the least objection to them; except it will be an

objection with you, that they are so much in your favour and in the favour of all our family.

I was speechless, absolutely speechless. Although my heart was ready to burst, yet could I neither weep nor speak.

I am sorry, said she, for your averseness to this match [*match* she was pleased to call it!]: but there is no help; and you must comply.

I was still speechless.

She folded the *warm statue,* as she was pleased to call me, in her arms; and entreated me, for Heaven's sake, and for her sake, to comply.

Speech and tears were lent me at the same time. You have given me life, madam, said I, clasping my uplifted hands together, and falling on one knee; a happy one till now has *your* goodness and my *papa's* made it! O do not, do not, make all the remainder of it miserable!

Your father, replied she, is resolved not to see you till he sees you as obedient a child as you used to be. This *is,* this *must* be, my last effort with you. Give me hope, my dear child: my peace is concerned.

To give you hope, my dearest, my most indulgent mamma, is to give you everything. Can I be honest, if I give a hope that I cannot confirm?

She was very angry. She again called me perverse: she upbraided me with regarding only my own prepossessions, and respecting not either her peace of mind or my own duty.

"I have had a very hard time of it, said she, between your father and you; for, seeing your dislike, I have more than once pleaded for you: but all to no purpose."

She went on: "Your father has declared that your unexpected opposition [*unexpected she was pleased to call it*], and Mr. Lovelace's continued menaces and insults, more and more convince him that a short day is necessary in order to put an end to all that man's hopes, and to his own apprehensions resulting from the disobedience of a child so favoured. He has, therefore, actually ordered patterns of the richest silks to be sent for from London——"

I started—I was out of breath—I gasped at this frightful precipitance. I was going to open with warmth against it. But she was pleased to hurry on, that I might not have time to express my disgusts at such a communication—to this effect:

"Your father, added she, at his going out, told me what he expected from me, in case I found that I had not the requisite influence upon you. It was this—that I should directly separate myself from you, and leave you singly to take the consequence of your double disobedience— I therefore entreat you, my dear Clarissa, concluded she, and that in the most earnest and condescending manner, to signify to your father, on his return, your ready obedience; and this as well for my sake as for your own."

Affected by my mother's goodness to me, and by that part of her argument which related to her own peace, I could not but wish it were possible for me to obey. I therefore paused, hesitated, considered, and was silent for some time. But then, recollecting that all was owing to the instigations of a brother and sister, wholly actuated by selfish and envious views; I would, madam said I, folding my hands, with an earnestness in which my whole heart was engaged, bear the cruellest tortures, bear loss of limb, and even of life to give *you* peace. But this man, every moment I would, at your command, think of him with favour, is the more my aversion. You cannot, indeed you cannot, think how my whole soul resists him! And to talk of contracts concluded upon; of patterns; of a short day! Save me, save me, O my dearest mamma, save your child from this heavy, from this insupportable evil!

Never was there a countenance that expressed so significantly, as my mother's did, an anguish, which she struggled to hide under an anger she was compelled to assume—till the latter overcoming the former, she turned from me with an uplifted eye, and stamping— *strange perverseness!* were the only words I heard of a sentence that she angrily pronounced; and was going. I then, half-franticly I believe, laid hold of her gown. Have patience with me, dearest madam! said I. Do not *you* renounce me totally! If you *must* separate yourself from your child, let it not be with *absolute* reprobation on *your own* part! My uncles may be hard-hearted—my father may be immovable. I may suffer from my brother's ambition, and from my sister's envy! But let me not lose my mamma's love; at least, her pity.

She turned to me with benigner rays. You *have* my love! You *have* my *pity!* But, O my dearest girl—I have not *yours.*

Indeed, indeed, madam, you have; and all my reverence, all my gratitude, you have! But in this *one* point: cannot I be this *once* obliged? Will no *expedient* be accepted? Have I not made a very fair proposal as to Mr. Lovelace?

I wish, for both our sakes, my dear unpersuadable girl, that the decision of this point lay with me. But why, when you know it does not, why should you thus perplex and urge me? To renounce Mr. Lovelace is now but *half* what is aimed at. Nor will anybody else believe you in earnest in the offer, if *I* would. While you remain single, Mr. Lovelace will have hopes—and you, in the opinion of others, inclinations.

Permit me, dearest madam, to say, that *your* goodness to me, *your* patience, *your* peace, weigh more with me than all the rest put together: for although I am to be treated by my brother, and, through his instigation, by my father, as a slave in this point, and not as a daughter, yet my mind is not that of a slave. You have brought me up to be mean.

So, Clary! you are already at defiance with your father! You forget that I must separate myself from you, if you will not comply. You do not remember that your father will take you up, where I leave you. Once more, however, I will put it to you: are you determined to brave your father's displeasure? Do you choose to break with us all, rather than encourage Mr. Solmes? Rather than give me hope?

Dreadful alternative! But is not my sincerity, is not the integrity of my heart, concerned in my answer? May not my everlasting happiness be the sacrifice? Forgive me, madam: bear with your child's boldness in such a cause as *this*! Settlements drawn! Patterns sent for! An early day! Dear, dear madam, how can I give hope, and not intend to be this man's?

Ah, girl, never say your *heart is free*! You deceive yourself if you think it is. You may guess what your father's first question on his return will be. He *must* know that I can do nothing with you. I have done my part. Seek *me,* if your mind change before he comes back: you have yet a little more time as he stays supper. I will no more seek *you* nor *to* you. And away she flung.

What could I do but weep?

Cl. H.

Miss Clarissa Harlowe to Miss Howe

Sat. Night.

I have been down.

I found my mother and sister together in my sister's parlour.

I entered like a dejected criminal; and besought the favour of a private audience.

You have, said she (looking at me with a sternness that never sits well on her sweet features), rather a *requesting* than a *conceding* countenance, Clarissa Harlowe: if I am mistaken, tell me so; and I will withdraw with you wherever you will. Yet, whether so or not, you may say what you have to say before your sister.

I come down, madam, said I, to beg of you to forgive me for anything you may have taken amiss in what passed above respecting your honoured self; and that you will be pleased to use your endeavours to soften my papa's displeasure against me on his return.

Such aggravating looks; such lifting up of hands and eyes; such a furrowed forehead in my sister!

My mother was angry enough without all that; and asked me to what purpose I came down, if I were still so untractable?

She had hardly spoken the words when Shorey came in to tell her that Mr. Solmes was in the hall, and desired admittance.

Ugly creature! I believe it was contrived, that he should be here at supper, to know the result of the conference between my mother and me, and that my father, on his return, might find us together.

I was hurrying away; but my mother commanded me (since I had come down only, as she said, to mock her) not to stir; and at the same time see if I could behave so to Mr. Solmes, as might encourage her to make the favourable report to my father which I had besought her to make.

My sister triumphed. I was vexed to be so caught, and to have such an angry and cutting rebuke given me.

The man stalked in. His usual walk is by pauses, as if he was telling his steps: and first paid his clumsy respects to my mother; then to my sister; next to me, as if I were already his wife, and therefore to be last in his notice; and sitting down by me, told us in general what weather it was. Then addressing himself to me; and how do *you* find it, miss? and would have taken my hand.

I withdrew it, I believe with disdain enough. My mother frowned. My sister bit her lip.

I could not contain myself: I never was so bold in my life; for I went on with my plea as if Mr. Solmes had not been there.

My mother coloured, and looked at him, at my sister, and at me.

My sister's eyes were opener and bigger than ever I saw them before.

The man understood me. He hemmed, and removed from one chair to another.

I went on, supplicating for my mother's favourable report: Nothing but invincible dislike, said I—

What would the girl be at, interrupted my mother? Why, Clary! Is this a subject! Is this!—is this!—is this a time—and again she looked upon Mr. Solmes.

I beg pardon, madam, said I. But my papa will soon return. And since I am not permitted to withdraw, it is not necessary, I humbly presume, that Mr. Solmes's presence should deprive me of this opportunity to implore your favourable report; and at the same time, if he still visit on my account (looking at him) to convince him, that it cannot possibly be to any purpose—

Is the girl mad? said my mother, interrupting me.

My sister, with the affectation of a whisper to my mother: This is—this is *spite,* madam [very *spitefully* she spoke the word] because you commanded her to stay.

I only looked at her, and turning to my mother, Permit me, madam, said I, to repeat my request. I have no brother, no sister! If I lose my mamma's favour I am lost for ever!

Mr. Solmes removed to his first seat, and fell to gnawing the head of his hazel; a carved head, almost as ugly as his own.

My sister rose, with a face all over scarlet, and stepping to the table, where lay a fan, she took it up, and although Mr. Solmes had observed that the weather was cold, fanned herself very violently.

My mother came to me, and angrily taking my hand, led me out of that parlour into my own; which, you know, is next to it. Is not this behaviour very bold, very provoking, think you, Clary?

I beg your pardon, madam, if it has that appearance to you. But indeed, my dear mamma, there seem to be snares laying for me. Too well I know my brother's drift.

My mother was about to leave me in high displeasure.

I besought her to stay: one favour, but one favour, dearest madam, said I, give me leave to beg of you—

What would the girl?

I see how everything is working about. I never, never can think of Mr. Solmes. My papa will be in tumults when he is told I cannot.

They will judge of the tenderness of your heart to a poor child who seems devoted by every one else, from the willingness you have already shown to hearken to my prayers. There will be endeavours used to confine me, and keep me out of your presence, and out of the presence of every one who used to love me [*this, my dear Miss* Howe, *is threatened*]. If this be effected; if it be put out of my power to plead my own cause, and to appeal to you, and to my Uncle Harlowe, of whom only I have hope; then will every ear be opened against me, and every tale encouraged—it is, therefore, my humble request, that, added to the disgraceful prohibitions I now suffer under, you will not, if you can help it, give way to my being denied *your* ear.

Your listening Hannah has given you this intelligence, as she does many others.

My Hannah, madam, listens not.

No more in Hannah's behalf—Hannah is known to make mischief—Hannah is known—but no more of that bold intermeddler. 'Tis true your father threatened to confine you to your chamber if you complied not, in order the more assuredly to deprive you of the opportunity of corresponding with those who harden your heart against his will. He bid me tell you so, when he went out, if I found you refractory. But I was loth to deliver so harsh a declaration; being still in hope that you would come down to us in a compliant temper. Hannah has overheard this, I suppose; as also that he declared he would break your heart rather than you should break his. And I now assure you, that you will be confined, and prohibited making teasing appeals to any of us: and we shall see who is to submit, you to us, or everybody to you.

And this, said I, is all I have to hope for from my mamma?

It is. But, Clary, this one further opportunity I give you: go in again to Mr. Solmes, and behave discreetly to him; and let your father find you together, upon *civil* terms at least.

My feet moved (of *themselves,* I think) farther from the parlour where he was, and towards the stairs; and there I stopped and paused.

If, proceeded she, you are determined to stand in defiance of us all—then indeed may you go up to your chamber (as you are ready to do)—and God help you!

God help me, indeed! for I cannot give hope of what I cannot intend.

I was moving to go up——

And *will* you go up, Clary?

I turned my face to her: my officious tears would needs plead for me: I could not just then speak; and stood still.

Good girl, distress me not thus! Dear, good girl, do not thus distress me! holding out her hand; but standing still likewise.

What *can* I do, madam? What *can* I do?

Go in again, my child. Go in again, my *dear* child! repeated she; and let your father find you together.

What, madam, to give *him* hope? To give hope to Mr. Solmes?

Obstinate, perverse, undutiful Clarissa! with a rejecting hand and angry aspect; then take your own way and go up!

She flung from me with high indignation: and I went up with a heavy heart; and feet as slow as my heart was heavy.

My father is come home, and my brother with him. Late as it is they are all shut up together.

The angry assembly is broken up. My two uncles and my Aunt Hervey are sent for, it seems, to be here in the morning to breakfast. I shall then, I suppose, know my doom. 'Tis past eleven, and I am ordered not to go to bed.

Twelve o'clock.

This moment the keys of everything are taken from me. It was proposed to send for me down: but my father said he could not bear to look upon me.

Cl. Harlowe.

Miss Clarissa Harlowe to Miss Howe

Sunday Morning, March 5.

Hannah has just brought me, from the private place in the garden wall, a letter from Mr. Lovelace, signed also by Lord M.

He tells me in it, "That Mr. Solmes makes it his boast that he is to be married in a few days to one of the shyest women in England: that my brother explains his meaning; this shy creature, he says, is me; and he assures every one that his younger sister is very soon to be Mr. Solmes's wife. He tells of the patterns bespoken which my mother mentioned to me."

Not one thing escapes him that is done or said in this house.

"He knows not, he says, what my relations' inducements can be to prefer such a man as Solmes to him. If advantageous settlements be the motive, Solmes shall not offer what he will refuse to comply with.

"As to his estate and family; the first cannot be excepted against: and for the second, he will not disgrace himself by a comparison so odious. He appeals to Lord M. for the regularity of his life and manners ever since he has made his addresses to me, or had hope of my favour.

"He desires my leave (in company with my lord, in a pacific manner) to attend my father or uncles, in order to make proposals that *must* be accepted, if they will but see him and hear what they are: and tells me, that he will submit to any measures that I shall prescribe, in order to bring about a reconciliation."

He presumes to be very earnest with me, "to give him a private meeting some night in my father's garden, attended by whom I please."

Really, my dear, were you to see his letter, you would think I had given him great encouragement, and that I am in direct treaty with him; or that he is sure that my friends will drive me into a foreign protection; for he has the boldness to offer, in my lord's name, an asylum to me should I be tyrannically treated in Solmes's behalf.

There are other particulars in this letter which I ought to mention to you: but I will take an opportunity to send you the letter itself, or a copy of it.

For my own part, I am very uneasy to think how I have been *drawn* on one hand, and *driven* on the other, into a clandestine, in short, into a mere lover-like correspondence which my heart condemns.

It is easy to see, if I do not break it off, that Mr. Lovelace's advantages, by reason of my unhappy situation, will every day increase, and I shall be more and more entangled. Yet if I do put an end to it, without making it a condition of being freed from Mr. Solmes's address.

All my relations are met. Mr. Solmes is expected. I am excessively uneasy. I must lay down my pen.

Sunday Noon.

What a cruel thing is suspense!

I desired to speak with Shorey. Shorey came. I directed her to carry to my mother my request for permission to go to church this afternoon. What think you was the return? Tell her that she must direct herself to her brother for any favour she has to ask.

I was resolved, however, to ask of *him* this favour. Accordingly, when they sent me up my solitary dinner, I gave the messenger a billet, in which I made it my humble request through him to my father, to be permitted to go to church this afternoon.

This was the contemptuous answer: "Tell her that her request will be taken into consideration *to-morrow*."

Patience will be the fittest return I can make to such an insult.

On recollection, I thought it best to renew my request. The following is a copy of what I wrote, and what follows that of the answer sent me.

Sir,—I know not what to make of the answer brought to my request of being permitted to go to church this afternoon. If you designed to show your pleasantry by it, I hope that will continue; and then my request will be granted.

My present situation is such that I never more wanted the benefit of the public prayers.

I will solemnly engage only to go thither and back again.

Nor will I, but by distant civilities, return the compliments of any of my acquaintance. My disgraces, if they are to have an end, need not to be proclaimed to the whole world.

> Your unhappy sister,
> Cl. Harlowe.

To Miss Clarissa Harlowe

For a girl to lay so much stress upon going to church, and yet resolve to defy her parents in an article of the greatest consequence to them, and to the whole family, is an absurdity. You are recommended, miss, to the practice of your *private* devotions. The *intention* is, I tell you plainly, to mortify you into a sense of your duty.

> Ja. Harlowe.

Monday Morning, March 6.

They are resolved to break my heart. My poor Hannah is discharged —disgracefully discharged! Thus it was:

Within half an hour after I had sent the poor girl down for my breakfast, that bold creature Betty Barnes, my sister's confident and servant (if a favourite maid and confident can be deemed a *servant*), came up.

What, miss, will you please to have for breakfast?

I was surprised. What will I have for breakfast, Betty! How!— what!—how comes it! Then I named Hannah. I could not tell what to say.

Don't be surprised, miss: but you'll see Hannah no more in this house.

God forbid! Is any harm come to Hannah? What! What is the matter with Hannah?

Why, miss, the short and the long is this: your papa and mamma think Hannah has stayed long enough in the house to do mischief; and so she is ordered to *troop* (that was the confident creature's word); and I am directed to wait upon you in her stead.

I burst into tears. I have no service for you, Betty Barnes: none at all. But where is Hannah? Cannot I speak with the poor girl? I owe her half a year's wages. I may never see her again perhaps; for they are resolved to break my heart.

And they think you are resolved to break theirs: so tit for tat, miss.

Impertinent I called her; and asked her if it were upon such confident terms that her service was to begin.

I was so very earnest to see the poor maid that (to *oblige* me, as she said) she went down with my request.

The worthy creature was as earnest to see me; and the favour was granted in presence of Shorey and Betty.

I thanked her when she came up for her past service to me.

Her heart was ready to break.

I gave her a little linen, some laces, and other odd things; and instead of four pounds which were due to her, ten guineas: and said, if ever I were again allowed to be my own mistress, I would think of *her* in the first place.

Hannah told me, before their faces, having no other opportunity, that she had been examined about letters *to* me and *from* me: and that she had given her pockets to Miss Harlowe, who looked into them, and put her fingers in her stays, to satisfy herself that she had not any.

We wept over each other at parting.

Miss Clarissa Harlowe to Miss Howe

Monday near 12 o'clock.

The enclosed letter was just now delivered to me. My brother has carried all his points.

Mond., March 6.

Miss Clary,—By command of your father and mother I write expressly to forbid you to come into their presence or into the garden when *they* are there: nor when they are *not* there, but with Betty Barnes to attend you; except by particular licence or command.

On their blessings, you are forbidden likewise to correspond with the vile Lovelace; as it is well known you did by means of your sly Hannah.

Neither are you to correspond with Miss Howe; who has given herself high airs of late; and might possibly help on your correspondence with that detested libertine. Nor, in short, with anybody without leave.

You are not to enter into the presence of either of your uncles without their leave first obtained. It is in *mercy* to you, after such a behaviour to your mother, that your father refuses to see you.

You are not to be seen in any apartment of the house you so lately governed as you pleased, unless you are commanded down.

In short, you are strictly to confine yourself to your chamber, except now and then in Betty Barnes's sight (as aforesaid) you take a morning or evening turn in the garden: and then you are to go directly, and without stopping at any apartment in the way, up and down the back stairs, that the sight of so perverse a young creature may not add to the pain you have given everybody.

The hourly threatenings of your fine fellow, as well as your own unheard-of obstinacy, will account to you for all this. What must your perverseness have been, that *such* a mother can give you up!

If anything I have written appear severe or harsh, it is still in your power (but perhaps will not always be so) to remedy it; and that by a single word.

Ja. Harlowe.

To JAMES HARLOWE, JUNIOR, ESQ.

Sir,—I will only say that you may congratulate yourself on having *so far* succeeded in all your views, that you may report what you please of me, and I can no more defend myself than if I were dead. Yet one favour, nevertheless, I will beg of you. It is this: that you will not occasion more severities, more disgraces, than are necessary for carrying into execution your further designs, whatever they be, against

Your unhappy sister,
Clarissa Harlowe.

MISS CLARISSA HARLOWE TO MISS HOWE

Tuesday, March 7.

Can such measures be supposed to soften? But surely they can only mean to try to frighten me into my brother's views! All my hope is to be able to weather this point till my Cousin Morden comes from Florence; and he is soon expected: yet, if they are determined upon a short day, I doubt he will not be here time enough to save me.

I asked Mrs. Betty if she had any orders to watch or attend me; or whether I was to ask *her* leave whenever I should be disposed to walk in the garden or to go to feed my bantams? Lord bless her! what could I mean by such a question!

However, as it behoved me to be assured on this head, I went down directly, and stayed an hour without question or impediment; and yet a good part of the time I walked under and in *sight,* as I may say, of my brother's study window, where both he and my sister happened to be. And I am sure they saw me, by the loud mirth they affected, by way of insult, as I suppose.

Tuesday Night.

Since I wrote the above, I ventured to send a letter by Shorey to my mother.

JEAN VALENTINE

Sex

All the years waiting, the whole, barren, young
Life long. The gummy yearning
All night long for the far white oval
Moving on the ceiling;
The hand on the head, the hand in hand;
The gummy pages of dirty books by flashlight,
Blank as those damaged classical groins;

Diffusion of leaves on the night sky,
The queer, sublunar walks.
And the words: the lily, the flame, the truelove knot,
Forget-me-not; coming, going,
Having, taking, lying with,
Knowing, dying;
The old king's polar sword,
The wine glass shattered on the stone floor.

And the thing itself not the thing itself,
But a metaphor.

Women in Love

Women in literature are often women in love—Anna Karenina, Emma Bovary, Phaedra, Ariadne, Mathilde de la Mole, Ursula Brangwen, the Princess of Cleves, Desdemona, Elizabeth Bennet (and the heroines of innumerable romances of somewhat lesser quality)—but I have bypassed the more celebrated lovers for a selection of five poems, a story, part of a novel. The poems are spoken in women's voices; in lyric love poetry by men we see the woman as the beloved, a reflection of her lover's passion. Not so with the poems of Adrienne Rich and Carol Bergé, where we follow a woman's discovery of herself as she explores the man loved, her love, herself as beloved. Bergé's poem is drawn from her An American Romance, a history of a love told in a relentless diary of poems that marks a stage in woman's awareness of her being. Rich's comes from a long sequence concluding Leaflets, the second of three collections of poetry (Necessities of Life was earlier, The Will to Change followed) that abandon the traditional forms and emotional evasions still common to women poets to follow the nerve-edged days of a mature woman living in her passions, haunted by the violence of war and city life, trying to find ways of keeping open the tracks to a self that refuses either to be overwhelmed by confusion or to turn away from complexity.

H. D.'s "Love That I Bear" is a characteristic poem, hard-edged and crystalline, distancing the emotion through precise language and imagery. Although known primarily as a poet and as one of the purest Imagists, H. D. was also the author of a little known novel, Palimpsest, composed of three long stories, each centering on a sensitive woman in a particular moment of time—Rome about 75 B.C., post–World War I London, and Egypt in the 1920s. In Bid Me to Live, a novel

written near the end of her life, H. D. draws on her marriage to Richard Aldington and her involvement with D. H. Lawrence to develop
the character of a woman artist who needs both love and independence.

CAROL BERGÉ
(1 9 2 8 –   )

Another View into Love

White flash—as the whole of it
springs sudden from visionary view
of yr femur—that longest bone
of all: immaculate crisp pale.
Freed of the loved expected flesh.
So that you stand near me stripped
of the ivory and scarlet uniform
in which you habitually greet me.
Your name is "skeleton"—a Greek
concept for this formal beauty . . .
your structure shines and shines,
you have become parthenon intact,
all your long fine bones sing out
perfect in symmetry, perfectly you.
A man—withal, the history of man
evident, in rib, knee, pelvic cup,
in set of skull on glossy knobs
of neck, and spine. That edifice!
I stare at sharpness, roundness,
places where it joins to move you.
The night, brilliant with yr bones,
spins and spins around us, as we
stand, man and woman, fleshed and
clothed, transfixed into this time.

It is the flesh that joins, when
loving. It is the bones remain
pristine, apart, and classical.

———————————

ADRIENNE RICH
(1929–   )

from Ghazals: Homage to Ghalib

A piece of thread ripped-out from a fierce design,
some weaving figured as magic against oppression.

I'm speaking to you as a woman to a man:
when your blood flows I want to hold you in my arms.

How did we get caught up fighting this forest fire,
we, who were only looking for a still place in the woods?

How frail we are, and yet, dispersed, always returning,
the barnacles they keep scraping from the warship's hull.

The hairs on your breast curl so lightly as you lie there,
while the strong heart goes on pounding in its sleep.

———————————

H. D.
(1 8 8 6 – 1 9 6 1)

Love That I Bear

Love that I bear
within my heart, O speak;
tell how beneath the serpent-spotted shell,
the cygnets wait,
how the soft owl
opens and flicks with pride,
eye-lids of great bird-eyes,
when underneath its breast,
the owlets shrink and turn.

———————————

Why did Emily Dickinson gradually become a recluse? The fact that no one knows has not prevented many from theorizing, and the usual bias has been in the conventional direction of disappointed love (for what else could govern a woman's life?). That seems entirely too easy for a woman whose poems and letters show her to be far from simple. Did she withdraw to find more time for her thoughts and words? That is more likely. And also possibly to escape the endless round of duties expected of the nineteenth-century housekeeper, duties undertaken by her sister Lavinia, another spinster whose life never moved far from childhood grooves.

Except for her poetic gift, Emily Dickinson was not unique, either in life or in literature, especially in the nineteenth century, when women were hedged by restrictions of custom and society to an even greater degree than in earlier times. One way of assertion, however, was left to them—that of renunciation. André Gide's *Alissa* in Strait Is the Gate is a wonderfully pure example of the Protestant poetry of renunciation, the scruples and self-dramatization which could give to an outwardly uneventful life a great inner drama. The Journals of the Swiss Protestant Henri Amiel and of the French Catholic Eugénie Guérin are two examples drawn from life (and what of Christina Rossetti and Alice James?); the entire drama is played out in the pages of journals where self-scrutiny and analysis, refined to an incredible degree, take the place of visible action.

Nor was this kind of self-sacrifice rare in Dickinson's New England, in her day already a region of dying farms and villages and of self-denying spinsters and bachelors who stepped aside to permit others to marry, who supported aging parents or worked the family farm, who displaced their affections onto nature, poetry, animals, or good works. Louisa May Alcott looked closely at her own financially strapped family life and never married. Hawthorne's sister-in-law Elizabeth Peabody (the model for Miss Birdseye in The Bostonians) was a celebrated do-gooder. What did marriage have to offer to an independent, intelligent woman but a houseful of children, incessant pregnancies, and unending chores? When the dazzling Julia Ward married the dashing Samuel Gridley Howe, it seemed an ideal match, but when Julia wished to add her voice to the chorus demanding female suffrage, she had to contend with a husband who had fought for the liberty of the Greeks but who refused to allow her to speak publicly for the rights of women. The latter half of that marriage was decidedly no model for a spirited woman to emulate. Quite possibly Dickinson's renunciation was simply the most likely choice in the circumstances.

Sarah Orne Jewett was another New England spinster, a doctor's daughter whose stories deal with the shrinking villages and declining economic life of coastal Maine at the end of the nineteeth century. Her characters are lonely men and women, the disappointed and isolated, people trying to make the best out of a life of pieces and scraps, heirs and heiresses of diminished fortunes (as in the sad little comedy "The Dulham Ladies"). Martha in "Martha's Lady" is the provincial cousin of Flaubert's Félicité (in "A Simple Heart"), observed by Jewett without Flaubert's formidable irony. In Martha's restricted life there has been only one outlet for her love and she bestows it freely on Helena Vernon without any awareness of the pathos of sustaining this love for *a lifetime on the basis of a summer's visit. For her it is a great love and it is enough. Martha stands for all the silent women "who did not tell [their] love," who stepped aside and who waited, not perhaps the love we most immediately think of, not the women in love of Lawrence, but no less a love in its own way.*

EMILY DICKINSON
(1830–1886)

I had been hungry all the years;
My noon had come, to dine;
I, trembling, drew the table near,
And touched the curious wine.

'Twas this on tables I had seen,
When turning, hungry, lone,
I looked in windows, for the wealth
I could not hope to own.

I did not know the ample bread,
'Twas so unlike the crumb
The birds and I had often shared
In Nature's dining-room.

The plenty hurt me, 'twas so new,—
Myself felt ill and odd,
As berry of a mountain bush
Transplanted to the road.

Nor was I hungry; so I found
That hunger was a way,
Of persons outside windows,
The entering takes away.

Wild nights! Wild nights!
Were I with thee,
Wild nights should be
Our luxury!

Futile the winds
To a heart in port,—
Done with the compass,
Done with the chart.

Rowing in Eden!
Ah! the sea!
Might I but moor
To-night in thee!

SARAH ORNE JEWETT

(1849–1909)

Martha's Lady

One day, many years ago, the old Judge Pyne house wore an unwonted look of gayety and youthfulness. The high-fenced green garden
was bright with June flowers. Under the elms in the large shady front
yard you might see some chairs placed near together, as they often used
to be when the family were all at home and life was going on gayly
with eager talk and pleasure-making; when the elder judge, the grandfather, used to quote that great author, Dr. Johnson, and say to his
girls, "Be brisk, be splendid, and be public."

One of the chairs had a crimson silk shawl thrown carelessly over its
straight back, and a passer-by, who looked in through the latticed gate
between the tall gate-posts with their white urns, might think that this
piece of shining East Indian color was a huge red lily that had suddenly bloomed against the syringa bush. There were certain windows

thrown wide open that were usually shut, and their curtains were blowing free in the light wind of a summer afternoon; it looked as if a large household had returned to the old house to fill the prim best rooms and find them full of cheer.

It was evident to everyone in town that Miss Harriet Pyne, to use the village phrase, had company. She was the last of her family, and was by no means old; but being the last, and wonted to live with people much older than herself, she had formed all the habits of a serious elderly person. Ladies of her age, something past thirty, often wore discreet caps in those days, especially if they were married, but being single, Miss Harriet clung to youth in this respect, making the one concession of keeping her waving chestnut hair as smooth and stiffly arranged as possible. She had been the dutiful companion of her father and mother in their latest years, all her elder brothers and sisters having married and gone, or died and gone, out of the old house. Now that she was left alone it seemed quite the best thing frankly to accept the fact of age, and to turn more resolutely than ever to the companionship of duty and serious books. She was more serious and given to routine than her elders themselves, as sometimes happened when the daughters of New England gentlefolks were brought up wholly in the society of their elders. At thirty-five she had more reluctance than her mother to face an unforeseen occasion, certainly more than her grandmother, who had preserved some cheerful inheritance of gayety and worldliness from colonial times.

There was something about the look of the crimson silk shawl in the front yard to make one suspect that the sober customs of the best house in a quiet New England village were all being set at defiance, and once when the mistress of the house came to stand in her own doorway, she wore the pleased but somewhat apprehensive look of a guest. In these days New England life held the necessity of much dignity and discretion of behavior; there was the truest hospitality and good cheer in all occasional festivities, but it was sometimes a self-conscious hospitality, followed by an inexorable return to asceticism both of diet and of behavior. Miss Harriet Pyne belonged to the very dullest days of New England, those which perhaps held the most priggishness for the learned professions, the most limited interpretation of the word "evangelical," and the pettiest indifference to large things. The outbreak of a desire for larger religious freedom caused at first a most determined reaction toward formalism, especially in small and quiet villages like Ashford, intently busy with their own concerns. It was high time for a little leaven to begin its work, in this moment when the great impulses of the war for liberty had died away and those of the coming war for patriotism and a new freedom had hardly yet begun.

The dull interior, the changed life of the old house, whose former activities seemed to have fallen sound asleep, really typified these larger conditions, and a little leaven had made its easily recognized appearance in the shape of a light-hearted girl. She was Miss Harriet's young Boston cousin, Helena Vernon, who, half-amused and half-impatient at the unnecessary sober-mindedness of her hostess and of Ashford in general, had set herself to the difficult task of gayety. Cousin Harriet looked on at a succession of ingenious and, on the whole, innocent attempts at pleasure, as she might have looked on at the frolics of a kitten who easily substitutes a ball of yarn for the uncertainties of a bird or a wind-blown leaf, and who may at any moment ravel the fringe of a sacred curtain-tassle in preference to either.

Helena, with her mischievous appealing eyes, with her enchanting old songs and her guitar, seemed the more delightful and even reasonable because she was so kind to everybody, and because she was a beauty. She had the gift of most charming manners. There was all the unconscious lovely ease and grace that had come with the good breeding of her city home, where many pleasant people came and went; she had no fear, one had almost said no respect, of the individual, and she did not need to think of herself. Cousin Harriet turned cold with apprehension when she saw the minister coming in at the front gate, and wondered in agony if Martha were properly attired to go to the door, and would by any chance hear the knocker; it was Helena who, delighted to have anything happen, ran to the door to welcome the Reverend Mr. Crofton as if he were a congenial friend of her own age. She could behave with more or less propriety during the stately first visit, and even contrive to lighten it with modest mirth, and to extort the confession that the guest had a tenor voice, though sadly out of practice; but when the minister departed a little flattered, and hoping that he had not expressed himself too strongly for a pastor upon the poems of Emerson, and feeling the unusual stir of gallantry in his proper heart, it was Helena who caught the honored hat of the late

Judge Pyne from its last resting-place in the hall, and holding it securely in both hands, mimicked the minister's self-conscious entrance. She copied his pompous and anxious expression in the dim parlor in such delicious fashion that Miss Harriet, who could not always extinguish a ready spark of the original sin of humor, laughed aloud.

"My dear!" she exclaimed severely the next moment, "I am ashamed of your being so disrespectful!" and then laughed again, and took the affecting old hat and carried it back to its place.

"I would not have had anyone else see you for the world," she said sorrowfully as she returned, feeling quite self-possessed again, to the parlor doorway; but Helena still sat in the minister's chair, with her small feet placed as his stiff boots had been, and a copy of his solemn expression before they came to speaking of Emerson and of the guitar. "I wish I had asked him if he would be so kind as to climb the cherry-tree," said Helena, unbending a little at the discovery that her cousin would consent to laugh no more. "There are all those ripe cherries on the top branches. I can climb as high as he, but I can't reach far enough from the last branch that will bear me. The minister is so long and thin—"

"I don't know what Mr. Crofton would have thought of you; he is a very serious young man," said cousin Harriet, still ashamed of her laughter. "Martha will get the cherries for you, or one of the men. I should not like to have Mr. Crofton think you were frivolous, a young lady of your opportunities"—but Helena had escaped through the hall and out at the garden door at the mention of Martha's name. Miss Harriet Pyne sighed anxiously, and then smiled, in spite of her deep convictions, as she shut the blinds and tried to make the house look solemn again.

The front door might be shut, but the garden door at the other end of the broad hall was wide open upon the large sunshiny garden, where the last of the red and white peonies and the golden lilies, and the first of the tall blue larkspurs lent their colors in generous fashion. The straight box borders were all in fresh and shining green of their new leaves, and there was a fragrance of the old garden's inmost life and soul blowing from the honeysuckle blossoms on a long trellis. It was now late in the afternoon, and the sun was low behind great apple-trees at the garden's end, which threw their shadows over the short turf of the bleaching-green. The cherry-trees stood at one side in full sunshine, and Miss Harriet, who presently came to the garden steps to watch like a hen at the water's edge, saw her cousin's pretty figure in its white dress of India muslin hurrying across the grass. She was accompanied by the tall, ungainly shape of Martha the new maid, who, dull and indifferent to everyone else, showed a surprising willingness and allegiance to the young guest.

"Martha ought to be in the dining-room, already, slow as she is; it wants but half an hour of tea-time," said Miss Harriet, as she turned and went into the shaded house. It was Martha's duty to wait at table, and there had been many trying scenes and defeated efforts toward her education. Martha was certainly very clumsy, and she seemed the clumsier because she had replaced her aunt, a most skillful person, who had but lately married a thriving farm and its prosperous owner. It must be confessed that Miss Harriet was a most bewildering instructor, and that her pupil's brain was easily confused and prone to blunders. The coming of Helena had been somewhat dreaded by reason of this incompetent service, but the guest took no notice of frowns or futile gestures at the first tea-table, except to establish friendly relations with Martha on her own account by a reassuring smile. They were about the same age, and next morning, before cousin Harriet came down, Helena showed by a word and a quick touch the right way to do something that had gone wrong and been impossible to understand the night before. A moment later the anxious mistress came in without suspicion, but Martha's eyes were as affectionate as a dog's, and there was a new look of hopefulness on her face; this dreaded guest was a friend after all, and not a foe come from proud Boston to confound her ignorance and patient efforts.

The two young creatures, mistress and maid, were hurrying across the bleaching-green.

"I can't reach the ripest cherries," explained Helena politely, "and I think Miss Pyne ought to send some to the minister. He has just made us a call. Why, Martha, you haven't been crying again!"

"Yes'm," said Martha sadly. "Miss Pyne always loves to send something to the minister," she acknowledged with interest, as if she did not wish to be asked to explain these latest tears.

"We'll arrange some of the best cherries in a pretty dish. I'll show you how, and you shall carry them over to the parsonage after tea," said Helena cheerfully, and Martha accepted the embassy with pleasure.

Life was beginning to hold moments of something like delight in the last few days.

"You'll spoil your pretty dress, Miss Helena," Martha gave shy warning, and Miss Helena stood back and held up her skirts with unusual care while the country girl, in her heavy blue checked gingham, began to climb the cherry-tree like a boy.

Down came the scarlet fruit like bright rain into the green grass.

"Break some nice twigs with the cherries and leaves together; oh, you're a duck, Martha!" and Martha, flushed with delight, and looking far more like a thin and solemn blue heron, came rustling down to earth again, and gathered the spoils into her clean apron.

That night at tea, during her handmaiden's temporary absence, Miss Harriet announced, as if by way of apology, that she thought Martha was beginning to understand something about her work. "Her aunt was a treasure, she never had to be told anything twice; but Martha has been as clumsy as a calf," said the precise mistress of the house. "I have been afraid sometimes that I never could teach her anything. I was quite ashamed to have you come just now, and find me so unprepared to entertain a visitor."

"Oh, Martha will learn fast enough because she cares so much," said the visitor eagerly. "I think she is a dear good girl. I do hope that she will never go away. I think she does things better every day, cousin Harriet," added Helena pleadingly, with all her kind young heart. The china-closet door was open a little way, and Martha heard every word. From that moment, she not only knew what love was like, but she knew love's dear ambitions. To have come from a stony hill-farm and a bare small wooden house, was like a cave-dweller's coming to make a permanent home in an art museum, such had seemed the elaborateness and elegance of Miss Pyne's fashion of life; and Martha's simple brain was slow enough in its processes and recognitions. But with this sympathetic ally and defender, this exquisite Miss Helena who believed in her, all difficulties appeared to vanish.

Later that evening, no longer homesick or hopeless, Martha returned from her polite errand to the minister, and stood with a sort of triumph before the two ladies, who were sitting in the front doorway, as if they were waiting for visitors, Helena still in her white muslin and red ribbons, and Miss Harriet in a thin black silk. Being happily self-forgetful in the greatness of the moment, Martha's manners were perfect, and she looked for once almost pretty and quite as young as she was.

"The minister came to the door himself, and returned his thanks. He said that cherries were always his favorite fruit, and he was much obliged to both Miss Pyne and Miss Vernon. He kept me waiting a few minutes, while he got this book ready to send to you, Miss Helena."

"What are you saying, Martha? I have sent him nothing!" exclaimed Miss Pyne, much astonished. "What does she mean, Helena?"

"Only a few cherries," explained Helena. "I thought Mr. Crofton would like them after his afternoon of parish calls. Martha and I arranged them before tea, and I sent them with our compliments."

"Oh, I am very glad you did," said Miss Harriet, wondering, but much relieved. "I was afraid—"

"No, it was none of my mischief," answered Helena daringly. "I did not think that Martha would be ready to go so soon. I should have shown you how pretty they looked among their green leaves. We put them in one of your best white dishes with the openwork edge. Martha shall show you to-morrow; mamma always likes to have them so." Helena's fingers were busy with the hard knot of a parcel.

"See this, cousin Harriet!" she announced proudly, as Martha disappeared round the corner of the house, beaming with the pleasures of adventure and success. "Look! the minister has sent me a book: Sermons on *what*? Sermons—it is so dark that I can't quite see."

"It must be his 'Sermons on the Seriousness of Life'; they are the only ones he has printed, I believe," said Miss Harriet, with much pleasure. "They are considered very fine discourses. He pays you a great compliment, my dear. I feared that he noticed your girlish levity."

"I behaved beautifully while he stayed," insisted Helena. "Ministers are only men," but she blushed with pleasure. It was certainly something to receive a book from its author, and such a tribute made her of more value to the whole reverent household. The minister was not only a man, but a bachelor, and Helena was at the age that best loves conquest; it was at any rate comfortable to be reinstated in cousin Harriet's good graces.

"Do ask the kind gentleman to tea! He needs a little cheering up," begged the siren in India muslin, as she laid the shiny black volume of sermons on the stone doorstep with an air of approval, but as if they had quite finished their mission.

"Perhaps I shall, if Martha improves as much as she has within the last day or two," Miss Harriet promised hopefully. "It is something I always dread a little when I am all alone, but I think Mr. Crofton likes to come. He converses so elegantly."

These were the days of long visits, before affectionate friends thought it quite worth while to take a hundred miles' journey merely to dine or to pass a night in one another's houses. Helena lingered through the pleasant weeks of early summer, and departed unwillingly at last to join her family at the White Hills, where they had gone, like other households of high social station, to pass the month of August out of town. The happy-hearted young guest left many lamenting friends behind her, and promised each that she would come back again next year. She left the minister a rejected lover, as well as the preceptor of the academy, but with their pride unwounded, and it may have been with wider outlooks upon the world and a less narrow sympathy both for their own work in life and for their neighbors' work and hindrances. Even Miss Harriet Pyne herself had lost some of the unnecessary provincialism and prejudice which had begun to harden a naturally good and open mind and affectionate heart. She was conscious of feeling younger and more free, and not so lonely. Nobody had ever been so gay, so fascinating, or so kind as Helena, so full of social resource, so simple and undemanding in her friendliness. The light of her young life cast no shadow on either young or old companions, her pretty clothes never seemed to make other girls look dull or out of fashion. When she went away up the street in Miss Harriet's carriage to take the slow train toward Boston and the gayeties of the new Profile House, where her mother waited impatiently with a group of Southern friends, it seemed as if there would never be any more picnics or parties in Ashford, and as if society had nothing left to do but to grow old and get ready for winter.

Martha came into Miss Helena's bedroom that last morning, and it was easy to see that she had been crying; she looked just as she did in that first sad week of homesickness and despair. All for love's sake she had been learning to do many things, and to do them exactly right; her eyes had grown quick to see the smallest chance for personal service. Nobody could be more humble and devoted; she looked years older than Helena, and wore already a touching air of caretaking.

"You spoil me, you dear Martha!" said Helena from the bed. "I don't know what they will say at home, I am so spoiled."

Martha went on opening the blinds to let in the brightness of the summer morning, but she did not speak.

"You are getting on splendidly, aren't you?" continued the little mistress. "You have tried so hard that you make me ashamed of myself. At first you crammed all the flowers together, and now you make them look beautiful. Last night cousin Harriet was so pleased when the table was so charming, and I told her that you did everything yourself, every bit. Won't you keep the flowers fresh and pretty in the house until I come back? It's so much pleasanter for Miss Pyne, and you'll feed my little sparrows, won't you? They're growing so tame."

"Oh, yes, Miss Helena!" and Martha looked almost angry for a moment, then she burst into tears and covered her face with her apron. "I couldn't understand a single thing when I first came. I never had been anywhere to see anything, and Miss Pyne frightened me when she talked. It was you made me think I could ever learn. I wanted to keep the place, 'count of mother and the little boys; we're dreadful hard pushed. Hepsy has been good in the kitchen; she said she ought to have patience with me, for she was awkward herself when she first came."

Helena laughed; she looked so pretty under the tasseled white curtains.

"I dare say Hepsy tells the truth," she said. "I wish you had told me about your mother. When I come again, some day we'll drive up country, as you call it, to see her. Martha I wish you would think of me sometimes after I go away. Won't you promise?" and the bright young face suddenly grew grave. "I have hard times myself; I don't always learn things that I ought to learn, I don't always put things straight. I wish you wouldn't forget me ever, and would just believe in me. I think it does help more than anything."

"I won't forget," said Martha slowly. "I shall think of you every day." She spoke almost with indifference, as if she had been asked to dust a room, but she turned aside quickly and pulled the little mat under the hot water jug quite out of its former straightness; then she hastened away down the long white entry, weeping as she went.

To lose out of sight the friend whom one has loved and lived to please is to lose joy out of life. But if love is true, there comes presently a higher joy of pleasing the ideal, that is to say, the perfect friend. The same old happiness is lifted to a higher level. As for Martha, the girl who stayed behind in Ashford, nobody's life could seem duller to those who could not understand; she was slow of step, and her eyes were almost always downcast as if intent upon incessant toil; but they startled you when she looked up, with their shining light. She was capable of the happiness of holding fast to a great sentiment, the ineffable satisfaction of trying to please one whom she truly loved. She never thought of trying to make other people pleased with herself; all she lived for was to do the best she could for others, and to conform to an ideal, which grew at last to be like a saint's vision, a heavenly figure painted upon the sky.

On Sunday afternoons in summer, Martha sat by the window of her chamber, a low-storied little room, which looked into the side yard and the great branches of an elm-tree. She never sat in the old wooden rocking-chair except on Sundays like this; it belonged to the day of rest and to happy meditation. She wore her plain black dress and a clean white apron, and held in her lap a little wooden box, with a brass ring on top for a handle. She was past sixty years of age and looked even older, but there was the same look on her face that it had sometimes worn in girlhood. She was the same Martha; her hands were old-looking and work-worn, but her face still shone. It seemed like yesterday that Helena Vernon had gone away, and it was more than forty years.

War and peace had brought their changes and great anxieties, the face of the earth was furrowed by floods and fire, the faces of mistress and maid were furrowed by smiles and tears, and in the sky the stars shone on as if nothing had happened. The village of Ashford added a few pages to its unexciting history, the minister preached, the people listened; now and then a funeral crept along the street, and now and then the bright face of a little child rose above the horizon of a family pew. Miss Harriet Pyne lived on in the large white house, which gained more and more distinction because it suffered no changes, save successive repaintings and a new railing about its stately roof. Miss Harriet herself had moved far beyond the uncertainties of an anxious youth. She had long ago made all her decisions, and settled all necessary questions; her scheme of life was as faultless as the miniature landscape of a Japanese garden, and as easily kept in order. The only important change she would ever be capable of making was the final change to another and a better world; and for that nature itself would gently provide, and her own innocent life.

Hardly any great social event had ruffled the easy current of life since Helena Vernon's marriage. To this Miss Pyne had gone, stately in appearance and carrying gifts of some old family silver which bore the Vernon crest, but not without some protest in her heart against the uncertainties of married life. Helena was so equal to a happy independence and even to the assistance of other lives grown strangely dependent upon her quick sympathies and instinctive decisions, that it was hard to let her sink her personality in the affairs of another. Yet a brilliant English match was not without its attractions to an old-fashioned gentlewoman like Miss Pyne, and Helena herself was amazingly happy; one day there had come a letter to Ashford, in which her very heart seemed to beat with love and self-forgetfulness, to tell cousin Harriet of such new happiness and high hope. "Tell Martha all that I say about my dear Jack," wrote the eager girl; "please show my letter to Martha, and tell her that I shall come home next summer and bring the handsomest and best man in the world to Ashford. I have told him all about the dear house and the dear garden; there never was such a lad to reach for cherries with his six-foot-two." Miss Pyne, wondering a little, gave the letter to Martha, who took it deliberately and as if she wondered too, and went away to read it slowly by herself. Martha cried over it, and felt a strange sense of loss and pain; it hurt her heart a little to read about the cherry-picking. Her idol seemed to be less her own since she had become the idol of a stranger. She never had taken such a letter in her hands before, but love at last prevailed, since Miss Helena was happy, and she kissed the last page where her name was written, feeling overbold, and laid the envelope on Miss Pyne's secretary without a word.

The most generous love cannot but long for reassurance, and Martha had the joy of being remembered. She was not forgotten when the day of the wedding drew near, but she never knew that Miss Helena had asked if cousin Harriet would not bring Martha to town; she should like to have Martha there to see her married. "She would help about the flowers," wrote the happy girl; "I know she will like to come, and I'll ask mamma to plan to have someone take her all about Boston and make her have a pleasant time after the hurry of the great day is over."

Cousin Harriet thought it was very kind and exactly like Helena, but Martha would be out of her element; it was most imprudent and girlish to have thought of such a thing. Helena's mother would be far from wishing for any unnecessary guest just then, in the busiest part of her household, and it was best not to speak of the invitation. Some day Martha should go to Boston if she did well, but not now. Helena did not forget to ask if Martha had come, and was astonished by the indifference of the answer. It was the first thing which reminded her that she was not a fairy princess having everything her own way in that last day before the wedding. She knew that Martha would have loved to be near, for she could not help understanding in that moment of her own happiness the love that was hidden in another heart. Next day this happy young princess, the bride, cut a piece of a great cake and put it into a pretty box that had held one of her wedding presents. With eager voices calling her, and all her friends about her, and her mother's face growing more and more wistful at the thought of parting, she still lingered and ran to take one or two trifles from her dressing-table, a little mirror and some tiny scissors that Martha would remember, and one of the pretty handkerchiefs marked with her maiden name. These she put in the box too; it was half a girlish freak and fancy, but she could not help trying to share her happiness, and Martha's life was so plain and dull. She whispered a message, and put the little package into cousin Harriet's hand for Martha as she said good-by. She was very fond of cousin Harriet. She smiled with a gleam of her old fun; Martha's puzzled look and tall awkward figure seemed to stand suddenly before her eyes, as she promised to come again to Ashford. Impatient voices called to Helena, her lover was at the door, and she hurried away, leaving her old home and her girlhood gladly. If she had only known it, as she kissed cousin Harriet good-by, they were never going to see each other again until they were old women. The first step that she took out of her father's house that day, married, and full of hope and joy, was a step that led her away from the green elms of Boston Common and away from her own country and those she loved best, to a brilliant, much-varied foreign life, and to nearly all the sorrows and nearly all the joys that the heart of one woman could hold or know.

On Sunday afternoons Martha used to sit by the window in Ashford and hold the wooden box which a favorite young brother, who after-

ward died at sea, had made for her, and she used to take out of it the
pretty little box with a gilded cover that had held the piece of wedding-
cake, and the small scissors, and the blurred bit of a mirror in its silver
case; as for the handkerchief with the narrow lace edge, once in two or
three years she sprinkled it as if it were a flower, and spread it out in
the sun on the old bleaching-green, and sat near by in the shrubbery
to watch lest some bold robin or cherry-bird should seize it and fly
away.

Miss Harriet Pyne was often congratulated upon the good fortune of
having such a helper and friend as Martha. As time went on this tall,
gaunt woman, always thin, always slow, gained a dignity of behavior
and simple affectionateness of look which suited the charm and dignity
of the ancient house. She was unconsciously beautiful like a saint, like
the picturesqueness of a lonely tree which lives to shelter unnumbered
lives and to stand quietly in its place. There was such rustic homeliness
and constancy belonging to her, such beautiful powers of apprehension,
such reticence, such gentleness for those who were troubled or sick; all
these gifts and graces Martha hid in her heart. She never joined the
church because she thought she was not good enough, but life was
such a passion and happiness of service that it was impossible not to be
devout, and she was always in her humble place on Sundays, in the
back pew next to the door. She had been educated by a remembrance;
Helena's young eyes forever looked at her reassuringly from a gay
girlish face. Helena's sweet patience in teaching her own awkwardness
could never be forgotten.

"I owe everything to Miss Helena," said Martha, half aloud, as she
sat alone by the window; she had said it to herself a thousand times.
When she looked in the little keepsake mirror she always hoped to see
some faint reflection of Helena Vernon, but there was only her own
brown old New England face to look back at her wonderingly.

Miss Pyne went less and less often to pay visits to her friends in
Boston; there were very few friends left to come to Ashford and make
long visits in the summer, and life grew more and more monotonous.
Now and then there came news from across the sea and messages of
remembrance, letters that were closely written on thin sheets of paper,
and that spoke of lords and ladies, of great journeys, of the death of
little children and the proud successes of boys at school, of the wedding

of Helena Dysart's only daughter; but even that had happened years ago. These things seemed far away and vague, as if they belonged to a story and not to life itself; the true links with the past were quite different. There was the unvarying flock of ground-sparrows that Helena had begun to feed; every morning Martha scattered crumbs for them from the side doorsteps while Miss Pyne watched from the dining-room window, and they were counted and cherished year by year.

Miss Pyne herself had many fixed habits, but little ideality or imagination, and so at last it was Martha who took thought for her mistress, and gave freedom to her own good taste. After a while, without anyone's observing the change, the every-day ways of doing things in the house came to be the stately ways that had once belonged only to the entertainment of guests. Happily both mistress and maid seized all possible chances for hospitality, yet Miss Harriet nearly always sat alone at her exquisitely served table with its fresh flowers, and the beautiful old china which Martha handled so lovingly that there was no good excuse for keeping it hidden on closet shelves. Every year when the old cherry-trees were in fruit, Martha carried the round white old English dish with a fretwork edge, full of pointed green leaves and scarlet cherries, to the minister, and his wife never quite understood why every year he blushed and looked so conscious of the pleasure, and thanked Martha as if he had received a very particular attention. There was no pretty suggestion toward the pursuit of the fine art of housekeeping in Martha's limited acquaintance with newspapers that she did not adopt; there was no refined old custom of the Pyne housekeeping that she consented to let go. And every day, as she had promised, she thought of Miss Helena,—oh, many times in every day: whether this thing would please her, or that be likely to fall in with her fancy or ideas of fitness. As far as was possible the rare news that reached Ashford through an occasional letter or the talk of guests was made part of Martha's own life, the history of her own heart. A worn old geography often stood open at the map of Europe on the lightstand in her room, and a little old-fashioned gilt button, set with a bit of glass like a ruby, that had broken and fallen from the trimming of one of Helena's dresses, was used to mark the city of her dwelling-place. In the changes of a diplomatic life Martha followed her lady all about the map. Sometimes the button was at Paris, and sometimes at Madrid;

once, to her great anxiety, it remained long at St. Petersburg. For such a slow scholar Martha was not unlearned at last, since everything about life in these foreign towns was of interest to her faithful heart. She satisfied her own mind as she threw crumbs to the tame sparrows; it was all part of the same thing and for the same affectionate reasons.

One Sunday afternoon in early summer Miss Harriet Pyne came hurrying along the entry that led to Martha's room and called two or three times before its inhabitant could reach the door. Miss Harriet looked unusually cheerful and excited, and she held something in her hand. "Where are you, Martha?" she called again. "Come quick, I have something to tell you!"

"Here I am, Miss Pyne," said Martha, who had only stopped to put her precious box in the drawer, and to shut the geography.

"Who do you think is coming this very night at half-past six? We must have everything as nice as we can; I must see Hannah at once. Do you remember my cousin Helena who has lived abroad so long? Miss Helena Vernon,—the Honorable Mrs. Dysart, she is now."

"Yes, I remember her," answered Martha, turning a little pale.

"I knew that she was in this country, and I had written to ask her to come for a long visit," continued Miss Harriet, who did not often explain things, even to Martha, though she was always conscientious about the kind messages that were sent back by grateful guests. "She telegraphs that she means to anticipate her visit by a few days and come to me at once. The heat is beginning in town, I suppose. I daresay, having been a foreigner so long, she does not mind traveling on Sunday. Do you think Hannah will be prepared? We must have tea a little later."

"Yes, Miss Harriet," said Martha. She wondered that she could speak as usual, there was such a ringing in her ears. "I shall have time to pick some fresh strawberries; Miss Helena is so fond of our strawberries."

"Why, I had forgotten," said Miss Pyne, a little puzzled by something quite unusual in Martha's face. "We must expect to find Mrs. Dysart a good deal changed, Martha; it is a great many years since she was here; I have not seen her since her wedding, and she has had a great deal of trouble, poor girl. You had better open the parlor chamber, and make it ready before you go down."

"It is all ready," said Martha. "I can carry some of those little sweet-brier roses upstairs before she comes."

"Yes, you are always thoughtful," said Miss Pyne, with unwonted feeling.

Martha did not answer. She glanced at the telegram wistfully. She had never really suspected before that Miss Pyne knew nothing of the love that had been in her heart all these years; it was half a pain and half a golden joy to keep such a secret; she could hardly bear this moment of surprise.

Presently the news gave wings to her willing feet. When Hannah, the cook, who never had known Miss Helena, went to the parlor an hour later on some errand to her old mistress, she discovered that this stranger guest must be a very important person. She had never seen the tea-table look exactly as it did that night, and in the parlor itself there were fresh blossoming boughs in the old East India jars, and lilies in the paneled hall, and flowers everywhere, as if there were some high festivity.

Miss Pyne sat by the window watching, in her best dress, looking stately and calm; she seldom went out now, and it was almost time for the carriage. Martha was just coming in from the garden with the strawberries, and with more flowers in her apron. It was a bright cool evening in June, the golden robins sang in the elms, and the sun was going down behind the apple-trees at the foot of the garden. The beautiful old house stood wide open to the long-expected guest.

"I think that I shall go down to the gate," said Miss Pyne, looking at Martha for approval, and Martha nodded and they went together slowly down the broad front walk.

There was a sound of horses and wheels on the roadside turf: Martha could not see at first; she stood back inside the gate behind the white lilac-bushes as the carriage came. Miss Pyne was there; she was holding out both arms and taking a tired, bent little figure in black to her heart. "Oh, my Miss Helena is an old woman like me!" and Martha gave a pitiful sob; she had never dreamed it would be like this; this was the one thing she could not bear.

"Where are you, Martha?" called Miss Pyne. "Martha will bring these in; you have not forgotten my good Martha, Helena?" Then Mrs. Dysart looked up and smiled just as she used to smile in the old days.

The young eyes were there still in the change face, and Miss Helena had come.

That night Martha waited in her lady's room just as she used to, humble and silent, and went through with the old unforgotten loving services. The long years seemed like days. At last she lingered a moment trying to think of something else that might be done, then she was going silently away, but Helena called her back. She suddenly knew the whole story and could hardly speak.

"Oh, my dear Martha!" she cried, "won't you kiss me good-night? Oh, Martha, have you remembered like this, all these long years!"

The Awakening, *in Edmund Wilson's words, "is a very odd book to have been written in America at the end of the nineteenth century . . . No case for free love or women's rights or the injustice of marriage is argued. The heroine is simply a sensuous woman who follows her inclinations without thinking much about these issues or tormenting herself with her conscience . . . [the novel is] quite uninhibited and beautifully written [and] anticipates D. H. Lawrence in its treatment of infidelity."* *

Edna Pontellier, a charming young woman from Kentucky, has been married for a few years to a moderately wealthy Creole businessman in New Orleans and is the mother of two small boys. She does not feel at all maternal, however, and fills her empty days with needlework, conversation, and entertaining, which fail to satisfy her. Summers spent on the Grande Isle in the Gulf of Mexico with other wives and families of the Creole aristocracy are long and lazy, with reading, gossip, and swimming the chief occupations. An outsider, Edna is not used to the Latin traditions of New Orleans, where married women of the utmost respectability engage in carefully circumscribed casual flirtations to relieve the tedium of long days at the beach. She falls in love with the young bachelor Robert Lebrun, and in the light of this sudden awakening the activities of her old life seem useless. Frightened by a passion so unacceptable in his society, Robert goes to Mexico on business.

* *Patriotic Gore* (New York: Oxford University Press, 1962), pp. 590–91.

Edna no longer cares to pretend that she loves her husband or chil-
dren, moves to a small house of her own, works at her drawing, and
tries to understand what has happened, but she lacks even a language
to describe her emotions. Nor does she have any sphere of action that
would permit her to discharge her energy in work. Not only her passion
but also her growing sense of herself as an independent person press
against the walls of her society until death becomes the only way out,
death from swimming out into the waters of the Gulf.

The following selection prefigures the conclusion of the short book
—Edna has finally learned to swim under the impetus of her newborn
love for Robert, and she follows this accomplishment with a direct
challenge to her husband.

The Awakening *lacks the density of Flaubert's* Madame Bovary *or*
Fontane's Effie Briest, *coming, as it does, from a provincial society, but*
it is a minor classic and retains its unique position in American litera-
ture as a study of a mature woman freeing herself from imposed re-
straints to taste the forbidden pleasures of sensuality and—even more
—of independence. "There was with her a feeling of having descended
in the social scale, with a corresponding sense of having risen in the
spiritual. Every step which she took toward relieving herself from obli-
gations added to her strength and expansion as an individual. She
began to look with her own eyes; to see and to apprehend the deeper
undercurrents of life." Because of her character, because of her society,
she chooses death over death-in-life.

KATE CHOPIN
(1851–1904)

The Awakening

Edna had attempted all summer to learn to swim. She had received
instructions from both the men and women; in some instances from
the children. Robert had pursued a system of lessons almost daily; and
he was nearly at the point of discouragement in realizing the futility
of his efforts. A certain ungovernable dread hung about her when in

the water, unless there was a hand near by that might reach out and reassure her.

But that night she was like the little tottering, stumbling, clutching child, who of a sudden realizes its powers, and walks for the first time alone, boldly and with over-confidence. She could have shouted for joy. She did shout for joy, as with a sweeping stroke or two she lifted her body to the surface of the water.

A feeling of exultation overtook her, as if some power of significant import had been given her to control the working of her body and her soul. She grew daring and reckless, overestimating her strength. She wanted to swim far out, where no woman had swum before.

Her unlooked-for achievement was the subject of wonder, applause, and admiration. Each one congratulated himself that his special teachings had accomplished this desired end.

"How easy it is!" she thought. "It is nothing," she said aloud; "why did I not discover before that it was nothing. Think of the time I have lost splashing about like a baby!" She would not join the groups in their sports and bouts, but intoxicated with her newly conquered power, she swam out alone.

She turned her face seaward to gather in an impression of space and solitude, which the vast expanse of water, meeting and melting with the moonlit sky, conveyed to her excited fancy. As she swam she seemed to be reaching out for the unlimited in which to lose herself.

Once she turned and looked toward the shore, toward the people she had left there. She had not gone any great distance—that is, what would have been a great distance for an experienced swimmer. But to her unaccustomed vision the stretch of water behind her assumed the aspect of a barrier which her unaided strength would never be able to overcome.

A quick vision of death smote her soul, and for a second of time appalled and enfeebled her senses. But by an effort she rallied her staggering faculties and managed to regain the land.

She made no mention of her encounter with death and her flash of terror, except to say to her husband, "I thought I should have perished out there alone."

"You were not so very far, my dear; I was watching you," he told her.

Edna went at once to the bath-house, and she had put on her dry clothes and was ready to return home before the others had left the water. She started to walk away alone. They all called to her and shouted to her. She waved a dissenting hand, and went on, paying no further heed to their renewed cries which sought to detain her.

"Sometimes I am tempted to think that Mrs. Pontellier is capricious," said Madame Lebrun, who was amusing herself immensely and feared that Edna's abrupt departure might put an end to the pleasure.

"I know she is," assented Mr. Pontellier; "sometimes, not often."

Edna had not traversed a quarter of the distance on her way home before she was overtaken by Robert.

"Did you think I was afraid?" she asked him, without a shade of annoyance.

"No; I knew you weren't afraid."

"Then why did you come? Why didn't you stay out there with the others?"

"I never thought of it."

"Thought of what?"

"Of anything. What difference does it make?"

"I'm very tired," she uttered, complainingly.

"I know you are."

"You don't know anything about it. Why should you know? I never was so exhausted in my life. But it isn't unpleasant. A thousand emotions have swept through me to-night. I don't comprehend half of them. Don't mind what I'm saying; I am just thinking aloud. I wonder if I shall ever be stirred again as Mademoiselle Reisz's playing moved me to-night. I wonder if any night on earth will ever again be like this one. It is like a night in a dream. The people about me are like some uncanny, half-human beings. There must be spirits abroad tonight."

"There are," whispered Robert. "Didn't you know this was the twenty-eighth of August?"

"The twenty-eighth of August?"

"Yes. On the twenty-eighth of August, at the hour of midnight, and if the moon is shining—the moon must be shining—a spirit that has haunted these shores for ages rises up from the Gulf. With its own penetrating vision the spirit seeks some one mortal worthy to hold him company, worthy of being exalted for a few hours into realms of the semi-celestials. His search has always hitherto been fruitless, and he has sunk back, disheartened, into the sea. But to-night he found Mrs. Pontellier. Perhaps he will never wholly release her from the spell. Perhaps she will never again suffer a poor, unworthy earthling to walk in the shadow of her divine presence."

"Don't banter me," she said, wounded at what appeared to be his flippancy. He did not mind the entreaty, but the tone with its delicate note of pathos was like a reproach. He could not explain; he could not tell her that he had penetrated her mood and understood. He said nothing except to offer her his arm, for, by her own admission, she was exhausted. She had been walking alone with her arms hanging limp, letting her white skirts trail along the dewy path. She took his arm, but she did not lean upon it. She let her hand lie listlessly, as though her thoughts were elsewhere—somewhere in advance of her body, and she was striving to overtake them.

Robert assisted her into the hammock which swung from the post before her door out to the trunk of a tree.

"Will you stay out here and wait for Mr. Pontellier?" he asked.

"I'll stay out here. Good-night."

"Shall I get you a pillow?"

"There's one here," she said, feeling about, for they were in the shadow.

"It must be soiled; the children have been tumbling it about."

"No matter." And having discovered the pillow, she adjusted it beneath her head. She extended herself in the hammock with a deep breath of relief. She was not a supercilious or an over-dainty woman. She was not much given to reclining in the hammock, and when she did so it was with no cat-like suggestion of voluptuous ease, but with a beneficent repose which seemed to invade her whole body.

"Shall I stay with you till Mr. Pontellier comes?" asked Robert, seating himself on the outer edge of one of the steps and taking hold of the hammock rope which was fastened to the post.

"If you wish. Don't swing the hammock. Will you get my white shawl which I left on the window-sill over at the house?"

"Are you chilly?"

"No; but I shall be presently."

"Presently?" he laughed. "Do you know what time it is? How long are you going to stay out here?"

"I don't know. Will you get the shawl?"

"Of course I will," he said, rising. He went over to the house, walking along the grass. She watched his figure pass in and out of the strips of moonlight. It was past midnight. It was very quiet.

When he returned with the shawl she took it and kept it in her hand. She did not put it around her.

"Did you say I should stay till Mr. Pontellier came back?"

"I said you might if you wished to."

He seated himself again and rolled a cigarette, which he smoked in silence. Neither did Mrs. Pontellier speak. No multitude of words could have been more significant than those moments of silence, or more pregnant with the first-felt throbbings of desire.

When the voices of the bathers were heard approaching, Robert said good-night. She did not answer him. He thought she was asleep. Again she watched his figure pass in and out of the strips of moonlight as he walked away.

"What are you doing out here, Edna? I thought I should find you in bed," said her husband, when he discovered her lying there. He had walked up with Madame Lebrun and left her at the house. His wife did not reply.

"Are you asleep?" he asked, bending down close to look at her.

"No." Her eyes gleamed bright and intense, with no sleepy shadows, as they looked into his.

"Do you know it is past one o'clock? Come on," and he mounted the steps and went into their room.

"Edna!" called Mr. Pontellier from within, after a few moments had gone by.

"Don't wait for me," she answered. He thrust his head through the door.

"You will take cold out there," he said, irritably. "What folly is this? Why don't you come in?"

"It isn't cold; I have my shawl."

"The mosquitoes will devour you."

"There are no mosquitoes."

She heard him moving about the room; every sound indicating impatience and irritation. Another time she would have gone in at his request. She would, through habit, have yielded to his desire; not with any sense of submission or obedience to his compelling wishes, but unthinkingly, as we walk, move, sit, stand, go through the daily tread-mill of the life which has been portioned out to us.

"Edna, dear, are you not coming in soon?" he asked again, this time fondly, with a note of entreaty.

"No; I am going to stay out here."

"This is more than folly," he blurted out. "I can't permit you to stay out there all night. You must come in the house instantly."

With a writhing motion she settled herself more securely in the hammock. She perceived that her will had blazed up, stubborn and resistant. She could not at that moment have done other than denied and resisted. She wondered if her husband had ever spoken to her like that before, and if she had submitted to his command. Of course she had; she remembered that she had. But she could not realize why or how she should have yielded, feeling as she then did.

"Léonce, go to bed," she said. "I mean to stay out here. I don't wish to go in, and I don't intend to. Don't speak to me like that again; I shall not answer you."

Mr. Pontellier had prepared for bed, but he slipped on an extra garment. He opened a bottle of wine, of which he kept a small and select supply in a buffet of his own. He drank a glass of the wine and went out on the gallery and offered a glass to his wife. She did not wish any. He drew up the rocker, hoisted his slippered feet on the rail, and proceeded to smoke a cigar. He smoked two cigars; then he went inside and drank another glass of wine. Mrs. Pontellier again declined to accept a glass when it was offered to her. Mr. Pontellier once more seated himself with elevated feet, and after a reasonable interval of time smoked some more cigars.

Edna began to feel like one who awakens gradually out of a dream, a delicious, grotesque, impossible dream, to feel again the realities pressing into her soul. The physical need for sleep began to overtake her; the exuberance which had sustained and exalted her spirit left her helpless and yielding to the conditions which crowded her in.

The stillest hour of the night had come, the hour before dawn, when the world seems to hold its breath. The moon hung low, and had turned from silver to copper in the sleeping sky. The old owl no longer hooted, and the water-oaks had ceased to moan as they bent their heads.

Edna arose, cramped from lying so long and still in the hammock.

She tottered up the steps, clutching feebly at the post before passing into the house.

"Are you coming in, Léonce?" she asked, turning her face toward her husband.

"Yes, dear," he answered, with a glance following a misty puff of smoke. "Just as soon as I have finished my cigar."

She slept but a few hours. They were troubled and feverish hours, disturbed with dreams that were intangible, that eluded her, leaving only an impression upon her half-awakened senses of something unattainable. She was up and dressed in the cool of the early morning. The air was invigorating and steadied somewhat her faculties. However, she was not seeking refreshment or help from any source, either external or from within. She was blindly following whatever impulse moved her, as if she had placed herself in alien hands for direction, and freed her soul of responsibility.

Most of the people at that early hour were still in bed and asleep. A few, who intended to go over to the *Chênière* for mass, were moving about. The lovers, who had laid their plans the night before, were already strolling toward the wharf. The lady in black, with her Sunday prayer-book, velvet and gold-clasped, and her Sunday silver beads, was following them at no great distance. Old Monsieur Farival was up, and was more than half inclined to do anything that suggested itself. He put on his big straw hat, and taking his umbrella from the stand in the hall, followed the lady in black, never overtaking her.

The little negro girl who worked Madame Lebrun's sewing-machine was sweeping the galleries with long, absent-minded strokes of the broom. Edna sent her up into the house to awaken Robert.

"Tell him I am going to the *Chênière*. The boat is ready; tell him to hurry."

He had soon joined her. She had never sent for him before. She had never asked for him. She had never seemed to want him before. She did not appear conscious that she had done anything unusual in commanding his presence. He was apparently equally unconscious of anything extraordinary in the situation. But his face was suffused with a quiet glow when he met her.

They went together back to the kitchen to drink coffee. There was no time to wait for any nicety of service. They stood outside the window and the cook passed them their coffee and a roll, which they drank and ate from the window-sill. Edna said it tasted good. She had not thought of coffee nor of anything. He told her he had often noticed that she lacked forethought.

"Wasn't it enough to think of going to the *Chênière* and waking you up?" she laughed. "Do I have to think of everything?—as Léonce says when he's in a bad humor. I don't blame him; he'd never be in a bad humor if it weren't for me."

They took a short cut across the sands. At a distance they could see the curious procession moving toward the wharf—the lovers, shoulder to shoulder, creeping; the lady in black, gaining steadily upon them; old Monsieur Farival, losing ground inch by inch, and a young bare-footed Spanish girl, with a red kerchief on her head and a basket on her arm, bringing up the rear.

Robert knew the girl, and he talked to her a little in the boat. No one present understood what they said. Her name was Mariequita. She had a round, sly, piquant face and pretty black eyes. Her hands were small, and she kept them folded over the handle of her basket. Her feet were broad and coarse. She did not strive to hide them. Edna looked at her feet, and noticed the sand and slime between her brown toes.

Beaudelet grumbled because Mariequita was there, taking up so much room. In reality he was annoyed at having old Monsieur Farival, who considered himself the better sailor of the two. But he would not quarrel with so old a man as Monsieur Farival, so he quarreled with Mariequita. The girl was deprecatory at one moment, appealing to Robert. She was saucy the next, moving her head up and down, making "eyes" at Robert and making "mouths" at Beaudelet.

The lovers were all alone. They saw nothing, they heard nothing. The lady in black was counting her beads for the third time. Old Monsieur Farival talked incessantly of what he knew about handling a boat, and of what Beaudelet did not know on the same subject.

Edna liked it all. She looked Mariequita up and down, from her ugly brown toes to her pretty black eyes, and back again.

"Why does she look at me like that?" inquired the girl of Robert.

"Maybe she thinks you are pretty. Shall I ask her?"

"No. Is she your sweetheart?"

"She's a married lady, and has two children."

"Oh! well! Francisco ran away with Sylvano's wife, who had four children. They took all his money and one of the children and stole his boat."

"Shut up!"

"Does she understand?"

"Oh, hush!"

"Are those two married over there—leaning on each other?"

"Of course not," laughed Robert.

"Of course not," echoed Mariequita, with a serious, confirmatory bob of the head.

The sun was high up and beginning to bite. The swift breeze seemed to Edna to bury the sting of it into the pores of her face and hands. Robert held his umbrella over her.

As they went cutting sidewise through the water, the sails bellied taut, with the wind filling and overflowing them. Old Monsieur Farival laughed sardonically at something as he looked at the sails, and Beaudelet swore at the old man under his breath.

Sailing across the bay to the *Chênière Caminada,* Edna felt as if she were being borne away from some anchorage which had held her fast, whose chains had been loosening—had snapped the night before when the mystic spirit was abroad, leaving her free to drift whithersoever she chose to set her sails. Robert spoke to her incessantly; he no longer noticed Mariequita. The girl had shrimps in her bamboo basket. They were covered with Spanish moss. She beat the moss down impatiently, and muttered to herself sullenly.

"Let us go to Grande Terre to-morrow?" said Robert in a low voice.

"What shall we do there?"

Independent Women

" 'Tis an evil lot to have a man's ambition and a woman's heart." Thus Margaret Fuller, who should have known, in nineteenth-century America. It is always to that century we return, the watershed, when more and more men moved out into the great world and more and more women retreated into the home to be hedged about with restrictions of custom and class. All that was lacking in the "respectable" classes was foot binding! Charlotte Brontë, for all her genius, traveled the weary route of the governess who lives on the fringes of other people's lives, but she was able to translate this demi-existence into art twice, in Jane Eyre and in the far better Villette, a brilliant, underrated book dense with the atmosphere of repressed sexual hysteria. Narrator and protagonist is poor Lucy Snowe, one of those respectable Victorian females with neither father nor husband as protector and so an acute embarrassment to a social order which made no provision for such ladies. Proudly English and Protestant, she nevertheless decides to seek work in Catholic Villette, Labassecour (read Brussels, Belgium), and there encounters Madame Beck, who hires Lucy to work in the school she heads. Today Madame Beck would be a business tycoon; a century ago she exercised her talents within the confines of a girls' school. What a remarkable woman she is, though, and how well Brontë has understood her!

The same atmosphere of repressed sexuality and hysteria surrounds Olive Chancellor in The Bostonians, although Henry James* gives his

_____

* James created so many memorable woman characters that choosing only one was difficult; however, his novels do not take easily to excerpting, and Isabel Archer, Milly Theale, Kate Croy, Madame de Vionnet, Maggie Verver, and the others need to be read at full stretch. The subject matter of The Bostonians

novel a tone of high comedy utterly foreign to Brontë. Louise Bogan
describes The Bostonians as the story of two young women—"Olive
Chancellor, of a certain position and means, 'a spinster, as Shelley was
a lyric poet or as the month of August is sultry,' living on the water
side of Charles Street; and Verena Tarrant, the daughter of a mesmer-
ist and all-round charlatan, who lives in a wooden cottage 'with a little
naked piazza,' on an unpaved 'place' in Cambridge. Verena has a 'gift'
—the gift of eloquence. She is able to move audiences, speaking
inanities in a voice that James compares to both silver and gold. . . .
Gotten up in a costume resembling that of a circus rider, she opens
her pretty mouth and exerts her fresh and genuine charm upon a variety of audiences in the cause of women's rights. Olive is, by contrast, a far more complicated character. A woman of distinction . . .
no fool . . . Olive is yet sterilized by an aridity of spirit, baffled by
genteel prejudices, and warped by a nervous constitution . . . a rather
terrifying resultant of Puritanism gone to seed, a female organism
driven by a masculine will, without the saving graces of masculine
intelligence or feminine tenderness and insight." *

There is an opposing center of force to Olive in Basil Ransom, her
cousin, a gallant young Mississippian who brings from the defeated
and devastated South the traditional view of women and who pits
himself against Olive for possession of Verena. In the chapter I have
included here, Olive takes steps to bind Verena to her and apparently
succeeds. She will lose in the end, however, for Verena and Basil,
both with banal souls, richly deserve each other and James is a realist,
but even in her misery and humiliation Olive remains—flawed, in-
complete, victimized—the most compelling character in the novel and
testimony to the psychic costs of being an independent woman in the
last century.

Not that there were many! From Jane Austen's heroines—still
relatively free in outlook, although still chained to a destiny of marriage or despicable spinsterhood—nineteenth-century fiction is a steady
chronicle of frustration and a narrowing of acceptable postures for
women until we come upon the simpering idiocies of the Young Girl

---

and its tone—so much lighter than the usual James—finally tipped the balance
in its favor.
* *A Poet's Alphabet* (New York, 1970), p. 243.

of Thackeray (dear Amelia Osborne!), Dickens, and Trollope, a type
so beautifully skewered in Henry James's The Awkward Age *in Aggie,*
kept a convent-bred innocent until she marries, at which time she
plunges wildly ahead to sample all the pleasures now open to her.

When Thomas Hardy tried to create the figure of an independent
woman in Sue Bridehead, the heroine of Jude the Obscure, the furious protest that greeted the book turned him away forever from novels
to the writing of poetry. And, yet, how independent is Sue, how lib-
erated from the constraints imposed upon "good women"? Not very.
Sexual frigidity is the key to her character, not freedom, and the seeds
of her final collapse into dreary religiosity are present from the start,
for her entire conception of freedom is rooted in the same sentimentality that lay behind the elevation of women to a pedestal and kept them
sheltered—and imprisoned—in the walls of their homes.

The words Hawthorne gave to Hester Prynne (another victim)
in The Scarlet Letter apply to Sue as well: "She assured them, too,
of her firm belief that, at some brighter period, when the world should
have grown ripe for it, in Heaven's own time, a new truth would be
revealed, in order to establish the whole relation between man and
woman on a surer ground of mutual happiness."

Because it needs to be read as a whole and is easily available in in-
expensive editions, I have not included pages from Jude the Obscure.
Interesting and valuable in itself, it makes an excellent introduction
to the work of Dorothy Richardson by allowing us to see the extent
of her achievement in creating Miriam Henderson and setting her on
a conscious struggle for independence that is remarkably contemporary, for Miriam is a truly original character, a young woman who
does not allow herself to become a sexual object and who refuses, in
fact, to become sidetracked by sex, either in marriage or out. Not
until one of the later parts of the many-volumed Pilgrimage does she
succumb (to a character based on H. G. Wells), and by then she has
already established her independence. Another quotation from Margaret Fuller gives the essential theme of this neglected masterpiece: "I
love best to be a woman, but womanhood at present is too straitly
bound to give me scope. At hours I live truly as a woman, at others
I stifle. . . . Men disappoint me so. I weary in this playground of
boys!"

Miriam's story is Dorothy Richardson's, the creator disappearing behind the vital figure of her character, through whom all the action is reflected. We are always in the present inside Miriam's head, following a stream-of-consciousness narrative that covers her adult life, from her comfortable girlhood with three sisters to the break-up of the home and Miriam's rejection of a marriage for convenience or support. Instead she chooses independence as a girl Friday for a group of dentists at one pound a week and an attic "room of her own," where she has just been installed in the section which follows.

Louise Bogan writes that "at first [Miriam's] reflection is exquisitely clear, with the senses of the female perceiver unblurred. Later, Miriam becomes more tendentious; the arguments multiply and the lines dividing pure creation and repetitive obsession begin to show. For there is no doubt that Richardson was obsessed, concerning what to her was the irreducible gap between the nature and motives of women and men. But there is also little doubt that her findings had truth in them; modern psychological insight has confirmed many. And Miriam, as part of her gradual enlightenment, is finally able to recognize and acknowledge her deep and compulsive psychic scars." *

Miriam's story belongs to all her sisters in the twentieth century who move out from the protected circle to seek wholeness as people; there is scarcely an insight achieved today that Miriam does not offer. I was eighteen when I heard Elizabeth Bowen speak of Richardson as the great unknown English novelist of the twentieth century and hunted up her books and was astonished at both their brilliance and their neglect. A new edition of Pilgrimage in 1967 makes available to us what is probably the single most important artistic document of twentieth-century woman.

* *Ibid.*, p. 344.

CHARLOTTE BRONTË
(1816–1853)

Villette

Being delivered into the charge of the maîtresse, I was led through a long narrow passage into a foreign kitchen, very clean but very strange. It seemed to contain no means of cooking—neither fireplace nor oven; I did not understand that the great black furnace which filled one corner was an efficient substitute for these. Surely pride was not already beginning its whispers in my heart; yet I felt a sense of relief when, instead of being left in the kitchen, as I half anticipated, I was led forward to a small inner room termed a "cabinet." A cook in a jacket, a short petticoat and sabots, brought my supper: to wit—some meat, nature unknown, served in an odd and acid, but pleasant sauce; some chopped potatoes, made savoury with I know not what—vinegar and sugar, I think: a tartine, or slice of bread and butter, and a baked pear. Being hungry, I ate and was grateful.

After the "Prière du Soir," Madame herself came to have another look at me. She desired me to follow her upstairs. Through a series of the queerest little dormitories—which, I heard afterwards, had once been nuns' cells: for the premises were in part of ancient date—and through the oratory—a long, low, gloomy room, where a crucifix hung, pale, against the wall, and two tapers kept dim vigils—she conducted me to an apartment where three children were asleep in three tiny beds. A heated stove made the air of this room oppressive; and, to mend matters, it was scented with an odour rather strong than delicate: a perfume, indeed, altogether surprising and unexpected under the circumstances, being like the combination of smoke with some spirituous essence—a smell, in short, of whisky.

Beside a table, on which flared the remnant of a candle guttering to waste in the socket, a coarse woman, heterogeneously clad in a broad striped showy silk dress and a stiff apron, sat in a chair fast asleep. To complete the picture, and leave no doubt as to the state of matters, a bottle and an empty glass stood at the sleeping beauty's elbow.

Madame contemplated this remarkable tableau with great calm; she neither smiled nor scowled; no impress of anger, disgust, or surprise ruffled the equality of her grave aspect; she did not even wake the woman. Serenely pointing to a fourth bed, she intimated that it was to be mine; then, having extinguished the candle and substituted for it a night-lamp, she glided through an inner door, which she left ajar— the entrance to her own chamber, a large, well-furnished apartment, as was discernible through the aperture.

My devotions that night were all thanksgiving. Strangely had I been led since morning—unexpectedly had I been provided for. Scarcely could I believe that not forty-eight hours had elapsed since I left London, under no other guardianship than that which protects the passenger-bird—with no prospects but the dubious cloud-tracery of hope.

I was a light sleeper; in the dead of night I suddenly awoke. All was hushed, but a white figure stood in the room—Madame in her night-dress. Moving without perceptible sound, she visited the three children in the three beds; she approached me: I feigned sleep, and she studied me long. A small pantomime ensued, curious enough. I dare say she sat a quarter of an hour on the edge of my bed, gazing at my face. She then drew nearer, bent close over me; slightly raised my cap, and turned back the border so as to expose my hair; she looked at my hand lying on the bedclothes. This done, she turned to the chair where my clothes lay: it was at the foot of the bed. Hearing her touch and lift them, I opened my eyes with precaution, for I own I felt curious to see how far her taste for research would lead her. It led her a good way: every article did she inspect. I divined her motive for this proceeding, viz., the wish to form from the garments a judgment respecting the wearer, her station, means, neatness, etc. The end was not bad, but the means were hardly fair or justifiable. In my dress was a pocket; she fairly turned it inside out: she counted the money in my purse; she opened a little memorandum-book, coolly perused its contents, and took from between the leaves a small plaited lock of Miss Marchmont's grey hair. To a bunch of three keys, being those of my trunk, desk, and workbox, she accorded special attention: with these, indeed, she withdrew a moment to her own room. I softly rose in my bed and followed her with my eye: these keys, reader, were not brought back till they had left on the toilet of the adjoining room the impress of their wards in wax. All being thus done decently and in order, my property was returned to its place, my clothes were carefully refolded. Of what nature were the conclusions deduced from this scrutiny? Were they favourable or otherwise? Vain question. Madame's face of stone (for of stone in its present night aspect it looked: it had been human, and, as I said before, motherly, in the salon) betrayed no response.

Her duty done—I felt that in her eyes this business was a duty—she rose, noiseless as a shadow: she moved towards her own chamber; at the door, she turned, fixing her eye on the heroine of the bottle, who still slept and loudly snored. Mrs. Svini (I presume this was Mrs. Svini, *Anglicè* or *Hibernicè* Sweeny)—Mrs. Sweeny's doom was in Madame Beck's eye—an immutable purpose that eye spoke: Madame's visitations for shortcomings might be slow, but they were sure. All this was very un-English: truly I was in a foreign land.

The morrow made me further acquainted with Mrs. Sweeny. It seems she had introduced herself to her present employer as an English lady in reduced circumstances: a native, indeed, of Middlesex, professing to speak the English tongue with the purest metropolitan accent. Madame—reliant on her own infallible expedients for finding out the truth in time—had a singular intrepidity in hiring service off-hand (as indeed seemed abundantly proved in my own case). She received Mrs. Sweeny as nursery-governess to her three children. I need hardly explain to the reader that this lady was in effect a native of Ireland; her station I do not pretend to fix: she boldly declared that she had "had the bringing-up of the son and daughter of a marquis." I think, myself, she might possibly have been a hanger-on, nurse, fosterer, or washerwoman, in some Irish family: she spoke a smothered tongue, curiously overlaid with mincing cockney inflections. By some means or other she had acquired, and now held in possession, a wardrobe of rather suspicious splendour—gowns of stiff and costly silk, fitting her indifferently, and apparently made for other proportions than those they now adorned; caps with real lace borders, and—the chief item in the inventory, the spell by which she struck a certain awe through the household, quelling the otherwise scornfully-disposed teachers and servants, and, so long as her broad shoulders *wore* the folds of that majestic drapery, even influencing Madame herself—*a real Indian shawl*—"un véritable Cachmire," as Madame Beck said, with unmixed reverence and amaze. I feel quite sure that without this "Cachmire" she would not have kept her footing in the pensionnat for two days: by virtue of it, and it only, she maintained the same a month.

But when Mrs. Sweeny knew that I was come to fill her shoes, then it was that she declared herself—then did she rise on Madame Beck in her full power—then come down on me with her concentrated weight. Madame bore this revelation and visitation so well, so stoically, that I for very shame could not support it otherwise than with composure. For one little moment Madame Beck absented herself from the room: ten minutes after, an agent of the police stood in the midst of us. Mrs. Sweeny and her effects were removed. Madame's brow had not been ruffled during the scene—her lips had not dropped one sharply-accented word.

This brisk little affair of the dismissal was all settled before breakfast: order to march given, policemen called, mutineer expelled, "chambre d'enfants" fumigated and cleansed, windows thrown open, and every trace of the accomplished Mrs. Sweeny—even to the fine essence and spiritual fragrance which gave token so subtle and so fatal of the head and front of her offending—was annihilated from the Rue Fossette: all this, I say, was done between the moment of Madame Beck's issuing like Aurora from her chamber and that in which she coolly sat down to pour out her first cup of coffee.

About noon, I was summoned to dress Madame. (It appeared my place was to be a hybrid between gouvernante and lady's-maid.) Till noon, she haunted the house in her wrapping-gown, shawl, and soundless slippers. How would the lady-chief of an English school approve this custom?

The dressing of her hair puzzled me; she had plenty of it: auburn, unmixed with grey: though she was forty years old. Seeing my embarrassment, she said, "You have not been a femme de chambre in your own country?" And taking the brush from my hand, and setting me aside, not ungently or disrespectfully, she arranged it herself. In performing other offices of the toilet, she half-directed, half-aided me, without the least display of temper or impatience. (N.B., that was the first and last time I was required to dress her. Henceforth, on Rosine, the portress, devolved that duty.)

When attired, Madame Beck appeared a personage of a figure rather short and stout, yet still graceful in its own peculiar way; that is, with the grace resulting from proportion of parts. Her complexion was fresh and sanguine, not too rubicund; her eye, blue and serene; her dark silk dress fitted her as a French sempstress alone can make a dress fit: she looked well, though a little bourgeoise; as bourgeoise, indeed, she was. I know not what of harmony pervaded her whole person; and yet her face offered contrast, too: its features were by no means such as are usually seen in conjunction with a complexion of such blended freshness and repose: their outline was stern: her forehead was high but narrow; it expressed capacity and some benevolence, but no expanse; nor did her peaceful yet watchful eye ever know the fire which is kindled in the heart or the softness which flows thence. Her mouth was hard: it could be a little grim; her lips were thin. For sensibility and genius, with all their tenderness and temerity, I felt somehow that Madame would be the right sort of Minos in petticoats.

In the long run, I found she was something else in petticoats too. Her name was Modeste Maria Beck, *née* Kint: it ought to have been Ignacia. She was a charitable woman, and did a great deal of good. There never was a mistress whose rule was milder. I was told that she never once remonstrated with the intolerable Mrs. Sweeny, despite her tipsiness, disorder, and general neglect; yet Mrs. Sweeny had to go the moment her departure became convenient. I was told, too, that neither masters nor teachers were found fault with in that establishment; yet both masters and teachers were often changed: they vanished and others filled their places, none could well explain how.

The establishment was both a pensionnat and an externat: the externes, or day-pupils, exceeded one hundred in number; the boarders were about a score. Madame must have possessed high administrative powers; she ruled all these, together with four teachers, eight masters, six servants, and three children, managing at the same time to perfection the pupils' parents and friends; and that without apparent effort; without bustle, fatigue, fever, or any symptom of undue excitement: occupied she always was—busy, rarely. It is true that Madame had her own system for managing and regulating this mass of machinery; and a very pretty system it was: the reader has seen a specimen of it in that small affair of turning my pocket inside out, and reading my private memoranda. "Surveillance," "espionage,"—these were her watchwords.

Still, Madame knew what honesty was, and liked it—that is, when it did not obtrude its clumsy scruples in the way of her will and interest. She had a respect for "Angleterre"; and as to "les Anglaises," she would have the women of no other country about her own children, if she could help it.

Often in the evening, after she had been plotting and counterplotting, spying and receiving the reports of spies all day, she would come up to my room—a trace of real weariness on her brow—and she would sit down and listen while the children said their little prayers to me in English: the Lord's Prayer, and the hymn beginning "Gentle Jesus," these little Catholics were permitted to repeat at my knee; and, when I had put them to bed, she would talk to me (I soon gained enough French to be able to understand, and even answer her) about England and Englishwomen, and the reasons for what she was pleased to term their superior intelligence, and more real and reliable probity. Very good sense she often showed; very sound opinions she often broached: she seemed to know that keeping girls in distrustful restraint, in blind ignorance, and under a surveillance that left them no moment and no corner for retirement, was not the best way to make them grow up honest and modest women; but she averred that ruinous consequences would ensue if any other method were tried with continental children: they were so accustomed to restraint, that relaxation, however guarded, would be misunderstood and fatally presumed on. She was sick, she would declare, of the means she had to use, but use them she must; and after discoursing, often with dignity and delicacy, to me, she would move away on her "souliers de silence," and glide ghost-like through the house, watching and spying everywhere, peering through every key-hole, listening behind every door.

After all, Madame's system was not bad—let me do her justice. Nothing could be better than all her arrangements for the physical well-being of her scholars. No minds were overtasked; the lessons were well distributed and made incomparably easy to the learner; there was a liberty of amusement, and a provision for exercise which kept the girls healthy; the food was abundant and good: neither pale nor puny faces were anywhere to be seen in the Rue Fossette. She never grudged a holiday; she allowed plenty of time for sleeping, dressing, washing, eating; her method in all these matters was easy, liberal, salutary, and rational: many an austere English schoolmistress would do vastly well to imitate her—and I believe many would be glad to do so, if exacting English parents would let them.

As Madame Beck ruled by espionage, she of course had her staff of spies: she perfectly knew the quality of the tools she used, and while she would not scruple to handle the dirtiest for a dirty occasion—flinging this sort from her like refuse rind, after the orange has been duly squeezed—I have known her fastidious in seeking pure metal for clean uses; and when once a bloodless and rustless instrument was found, she was careful of the prize, keeping it in silk and cotton-wool. Yet, woe be to that man or woman who relied on her one inch beyond the point where it was her interest to be trustworthy: interest was the master-key of Madame's nature—the mainspring of her motives—the alpha and omega of her life. I have seen her *feelings* appealed to, and I have smiled in half-pity, half-scorn at the appellants. None ever gained her ear through that channel, or swayed her purpose by that means. On the contrary, to attempt to touch her heart was the surest way to rouse her antipathy, and to make of her a secret foe. It proved to her that she had no heart to be touched: it reminded her where she was impotent and dead. Never was the distinction between charity and mercy better exemplified than in her. While devoid of sympathy, she had a sufficiency of rational benevolence: she would give in the readiest manner to people she had never seen—rather, however, to classes than to individuals. "Pour les pauvres" she opened her purse freely—against *the poor man,* as a rule, she kept it closed. In philanthropic schemes for the benefit of society at large she took a cheerful part; no private sorrow touched her: no force or mass of suffering concentrated in one heart had power to pierce hers. Not the agony in Gethsemane, not the death on Calvary, could have wrung from her eyes one tear.

I say again, Madame was a very great and a very capable woman. That school offered her for her powers too limited a sphere; she ought to have swayed a nation: she should have been the leader of a turbulent legislative assembly. Nobody could have browbeaten her, none irritated her nerves, exhausted her patience, or over-reached her astuteness. In her own single person, she could have comprised the duties of a first minister and a superintendent of police. Wise, firm, faithless; secret, crafty, passionless; watchful and inscrutable; acute and insensate—withal perfectly decorous—what more could be desired?

The sensible reader will not suppose that I gained all the knowledge here condensed for his benefit in one month, or in one half-year. No! what I saw at first was the thriving outside of a large and flourishing educational establishment. Here was a great house, full of healthy, lively girls, all well-dressed and many of them handsome, gaining knowledge by a marvellously easy method, without painful exertion or useless waste of spirits; not, perhaps, making very rapid progress in anything; taking it easy, but still always employed, and never oppressed. Here was a corps of teachers and masters more stringently tasked, as all the real head-labour was to be done by them, in order to save the pupils, yet having their duties so arranged that they relieved each other in quick succession whenever the work was severe: here, in short, was a foreign school; of which the life, movement, and variety made it a complete and most charming contrast to many English institutions of the same kind.

Behind the house was a large garden, and in summer the pupils almost lived out of doors amongst the rose-bushes and the fruit-trees. Under the vast and vine-draped berceau, Madame would take her seat on summer afternoons, and send for the classes, in turns, to sit round her and sew and read. Meantime, masters came and went, delivering short and lively lectures, rather than lessons, and the pupils made notes of their instructions, or did *not* make them—just as inclination prompted; secure that, in case of neglect, they could copy the notes of their companions. Besides the regular monthly *jours de sortie,* the Catholic fête-days brought a succession of holidays all the year round; and sometimes on a bright summer morning, or soft summer evening, the boarders were taken out for a long walk into the country, regaled with *gaufres* and *vin blanc,* or new milk and *pain bis,* or *pistolets au beurre* (rolls) and coffee. All this seemed very pleasant, and Madame appeared goodness itself; and the teachers not so bad but they might be worse; and the pupils perhaps a little noisy and rough, but types of health and glee.

Thus did the view appear, seen through the enchantment of distance; but there came a time when distance was to melt for me—when I was to be called down from my watch-tower of the nursery, whence I had hitherto made my observations, and was to be compelled into closer intercourse with this little world of the Rue Fossette.

I was one day sitting upstairs, as usual, hearing the children their

English lessons, and at the same time turning a silk dress for Madame, when she came sauntering into the room with that absorbed air and brow of hard thought she sometimes wore, and which made her look so little genial. Dropping into a seat opposite mine, she remained some minutes silent. Désirée, the eldest girl, was reading to me some little essay of Mrs. Barbauld's, and I was making her translate currently from English to French as she proceeded, by way of ascertaining that she comprehended what she read: Madame listened.

Presently, without preface or prelude, she said, almost in the tone of one making an accusation, "Meess, in England you were a governess."

"No, Madame," said I, smiling, "you are mistaken."

"Is this your first essay at teaching—this attempt with my children?"

I assured her it was. Again she became silent; but looking up, as I took a pin from the cushion, I found myself an object of study: she held me under her eye; she seemed turning me round in her thoughts —measuring my fitness for a purpose, weighing my value in a plan. Madame had, ere this, scrutinised all I had, and I believe she esteemed herself cognisant of much that I was; but from that day, for the space of about a fortnight, she tried me by new tests. She listened at the nursery door when I was shut in with the children; she followed me at a cautious distance when I walked out with them, stealing within ear-shot whenever the trees of park or boulevard afforded a sufficient screen: a strict preliminary process having thus been observed, she made a move forward.

One morning, coming on me abruptly, and with the semblance of hurry, she said she found herself placed in a little dilemma; Mr. Wilson, the English master, had failed to come at his hour; she feared he was ill; the pupils were waiting in classe; there was no one to give a lesson; should I, for once, object to giving a short dictation exercise, just that the pupils might not have it to say they had missed their English lesson?

"In classe, Madame?" I asked.

"Yes, in classe; in the second division."

"Where there are sixty pupils," said I; for I knew the number, and with my usual base habit of cowardice I shrank into my sloth like a snail into a shell, and alleged incapacity and impracticability as a pretext to escape action. If left to myself, I should infallibly have let this chance slip. Inadventurous, unstirred by impulses of practical ambition, I was capable of sitting twenty years teaching infants the hornbook, turning silk dresses, and making children's frocks. Not that true contentment dignified this infatuated resignation: my work had neither charm for my taste, nor hold on my interest; but it seemed to me a great thing to be without heavy anxiety, and relieved from intimate trial: the negation of severe suffering was the nearest approach to happiness I expected to know. Besides, I seemed to hold two lives— the life of thought, and that of reality; and, provided the former was nourished with a sufficiency of the strange necromantic joys of fancy, the privileges of the latter might remain limited to daily bread, hourly work, and a roof of shelter.

"Come," said Madame, as I stooped more busily than ever over the cutting-out of a child's pinafore, "leave that work."

"But Fifine wants it, Madame."

"Fifine must want it, then, for I want *you*."

And as Madame Beck did really want and was resolved to have me—as she had long been dissatisfied with the English master, with his shortcomings in punctuality, and his careless method of tuition— as, too, *she* did not lack resolution and practical activity, whether *I* lacked them or not—she, without more ado, made me relinquish thimble and needle; my hand was taken into hers, and I was conducted downstairs. When we reached the carré, a large square hall between the dwelling-house and the pensionnat, she paused, dropped my hand, faced and scrutinised me. I was flushed, and tremulous from head to foot: tell it not in Gath, I believe I was crying. In fact, the difficulties before me were far from being wholly imaginary; some of them were real enough; and not the least substantial lay in my want of mastery over the medium through which I should be obliged to teach. I had, indeed, studied French closely since my arrival in Villette; learning its practice by day, and its theory in every leisure moment at night, to as late an hour as the rule of the house would allow candle-light; but I was far from yet being able to trust my powers of correct oral expression.

"Dites donc," said Madame, sternly, "vous sentez vous réellement trop faible?"

I might have said "Yes," and gone back to nursery obscurity, and there, perhaps, mouldered for the rest of my life; but looking up at

Madame, I saw in her countenance a something that made me think twice ere I decided. At that instant she did not wear a woman's aspect, but rather a man's. Power of a particular kind strongly limned itself in all her traits, and that power was not *my* kind of power: neither sympathy, nor congeniality, nor submission, were the emotions it awakened. I stood—not soothed, nor won, nor overwhelmed. It seemed as if a challenge of strength between opposing gifts was given, and I suddenly felt all the dishonour of my diffidence, all the pusillanimity of my slackness to aspire.

"Will you," she said, "go backward or forward?" indicating with her hand, first, the small door of communication with the dwelling-house, and then the great double portals of the classes or schoolrooms.

"En avant," I said.

"But," pursued she, cooling as I warmed, and continuing the hard look, from very antipathy to which I drew strength and determination, "can you face the classes, or are you over-excited?"

She sneered slightly in saying this: nervous excitability was not much to Madame's taste.

"I am no more excited than this stone," I said, tapping the flag with my toe: "or than you," I added, returning her look.

"Bon! But let me tell you these are not quiet, decorous English girls you are going to encounter. Ce sont des Labassecouriennes, rondes, franches, brusques, et tant soit peu rebelles."

I said: "I know; and I know, too, that though I have studied French hard since I came here, yet I still speak it with far too much hesitation—too little accuracy to be able to command their respect: I shall make blunders that will lay me open to the scorn of the most ignorant. Still, I mean to give the lesson."

"They always throw over timid teachers," said she.

"I know that too, Madame; I have heard how they rebelled against and persecuted Miss Turner"—a poor friendless English teacher, whom Madame had employed, and lightly discarded; and to whose piteous history I was no stranger.

"C'est vrai," said she coolly. "Miss Turner had no more command over them than a servant from the kitchen would have had. She was weak and wavering; she had neither tact nor intelligence, decision nor dignity. Miss Turner would not do for these girls at all."

I made no reply, but advanced to the closed schoolroom door.

"You will not expect aid from me, or from anyone," said Madame. "That would at once set you down as incompetent for your office."

I opened the door, let her pass with courtesy, and followed her. There were three schoolrooms, all large. That dedicated to the second division, where I was to figure, was considerably the largest, and accommodated an assemblage more numerous, more turbulent, and infinitely more unmanageable than the other two. In after days, when I knew the ground better, I used to think sometimes (if such a comparison may be permitted) that the quiet, polished, tame first division was to the robust, riotous, demonstrative second division what the English House of Lords is to the House of Commons.

The first glance informed me that many of the pupils were more than girls—quite young women; I knew that some of them were of noble family (as nobility goes in Labassecour), and I was well convinced that not one amongst them was ignorant of my position in Madame's household. As I mounted the estrade (a low platform, raised a step above the flooring), where stood the teacher's chair and desk, I beheld opposite to me a row of eyes and brows that threatened stormy weather—eyes full of insolent light, and brows hard and unblushing as marble. The continental "female" is quite a different being to the insular "female" of the same age and class: I never saw such eyes and brows in England. Madame Beck introduced me in one cool phrase, sailed from the room, and left me alone in my glory.

I shall never forget that first lesson, nor all the under-current of life and character it opened up to me. Then first did I begin rightly to see the wide difference that lies between the novelist's and poet's ideal "jeune fille" and the said "jeune fille" as she really is.

It seems that three titled belles in the first row had sat down predetermined that a bonne d'enfants should not give them lessons in English. They knew they had succeeded in expelling obnoxious teachers before now; they knew that Madame would at any time throw overboard a professeur or maîtresse who became unpopular with the school—that she never assisted a weak official to retain his place—that if he had not strength to fight, or tact to win his way, down he went: looking at "Miss Snowe," they promised themselves an easy victory.

Mesdemoiselles Blanche, Virginie, and Angélique opened the campaign by a series of titterings and whisperings; these soon swelled into murmurs and short laughs, which the remoter benches caught up and echoed more loudly. This growing revolt of sixty against one soon became oppressive enough; my command of French being so limited, and exercised under such cruel constraint.

Could I have but spoken in my own tongue, I felt as if I might have gained a hearing; for, in the first place, though I knew I looked a poor creature, and in many respects actually was so, yet nature had given me a voice that could make itself heard, if lifted in excitement or deepened by emotion. In the second place, while I had no flow, only a hesitating trickle of language, in ordinary circumstances, yet—under stimulus such as was now rife through the mutinous mass—I could, in English, have rolled out readily phrases stigmatising their proceedings as such proceedings deserved to be stigmatised; and then with some sarcasm, flavoured with contemptuous bitterness for the ringleaders, and relieved with easy banter for the weaker but less knavish followers, it seemed to me that one might possibly get command over this wild herd and bring them into training, at least. All I could now do was to walk up to Blanche—Mademoiselle de Melcy, a young baronne—the eldest, tallest, handsomest, and most vicious—stand before her desk, take from under her hand her exercise-book, remount the estrade, deliberately read the composition, which I found very stupid, and as deliberately, and in the face of the whole school, tear the blotted page in two.

This action availed to draw attention and check noise. One girl alone, quite in the background, persevered in the riot with undiminished energy. I looked at her attentively. She had a pale face, hair like night, broad strong eyebrows, decided features, and a dark, mutinous, sinister eye: I noted that she sat close by a little door, which door, I was well aware, opened into a small closet where books were kept. She was standing up for the purpose of conducting her clamour with freer energies. I measured her stature and calculated her strength. She seemed both tall and wiry; but, if the conflict were brief and the attack unexpected, I thought I might manage her.

Advancing up the room, looking as cool and careless as I possibly could, in short, *ayant l'air de rien,* I slightly pushed the door and found it was ajar. In an instant, and with sharpness, I had turned on her. In another instant she occupied the closet, the door was shut, and the key in my pocket.

It so happened that this girl, Dolores by name, and a Catalonian by race, was the sort of character at once dreaded and hated by all her associates; the act of summary justice above noted proved popular: there was not one present but, in her heart, liked to see it done. They were stilled for a moment; then a smile—not a laugh—passed from desk to desk: then—when I had gravely and tranquilly returned to the estrade, courteously requested silence, and commenced a dictation as if nothing at all had happened—the pens travelled peacefully over the pages, and the remainder of the lesson passed in order and industry.

"C'est bien," said Madame Beck, when I came out of class, hot and a little exhausted. "Ça ira."

She had been listening and peeping through a spy-hole the whole time.

From that day I ceased to be nursery governess, and became English teacher. Madame raised my salary; but she got thrice the work out of me she had extracted from Mr. Wilson, at half the expense.

HENRY JAMES
(1843–1916)

The Bostonians

She hoped she should not soon see him again, and there appeared to be no reason she should, if their intercourse was to be conducted by means of checks. The understanding with Verena was, of course, complete; she had promised to stay with her friend as long as her friend should require it. She had said at first that she couldn't give up her mother, but she had been made to feel that there was no question of giving up. She should be as free as air, to go and come; she could spend hours and days with her mother, whenever Mrs. Tarrant required her attention; all that Olive asked of her was that, for the time, she should regard Charles Street as her home. There was no struggle about this, for the simple reason that by the time the question came to the front Verena was completely under the charm. The idea of Olive's charm will perhaps make the reader smile; but I use the word not in its derived, but in its literal sense. The fine web of authority, of dependence, that her strenuous companion had woven about her, was now as dense as a suit of golden mail; and Verena was thoroughly interested in their great undertaking; she saw it in the light of an active, enthusiastic faith. The benefit that her father desired for her was now assured; she expanded, developed, on the most liberal scale. Olive saw the difference, and you may imagine how she rejoiced in it; she had never known a greater pleasure. Verena's former attitude had been girlish submission, grateful, curious sympathy. She had given herself, in her young, amused surprise, because Olive's stronger will and the incisive proceedings with which she pointed her purpose drew her on. Besides, she was held by hospitality, the vision of new social horizons, the sense of novelty, and the love of change. But now the girl was disinterestedly attached to the precious things they were to do together; she cared about them for themselves, believed in them ardently, had them constantly in mind. Her share in the union of the two young women was no longer passive, purely appreciative; it was passionate, too, and it put forth a beautiful energy. If Olive desired to get Verena into training, she could flatter herself that the process had already begun, and that her colleague enjoyed it almost as much as she. Therefore she could say to herself, without the imputation of heartlessness, that when she left her mother it was for a noble, a sacred use. In point of fact, she left her very little, and she spent hours in jingling, aching, jostled journeys between Charles Street and the stale suburban cottage. Mrs. Tarrant sighed and grimaced, wrapped herself more than ever in her mantle, said she didn't know as she was fit to struggle alone, and that, half the time, if Verena was away, she wouldn't have the nerve to answer the doorbell; she was incapable, of course, of neglecting such an opportunity to posture as one who paid with her heart's blood for leading the van of human progress. But Verena had an inner sense (she judged her mother now, a little, for the first time), that she would be sorry to be taken at her word, and that she felt safe enough in trusting to her daughter's generosity. She could not divest herself of the faith—even now that Mrs. Luna was gone, leaving no trace, and the gray walls of a sedentary winter were apparently closing about the two young women—she could not renounce the theory that a residence in Charles Street must at least produce some contact with the brilliant classes. She was vexed at her daughter's resignation to not going to parties and to Miss Chancellor's not giving them; but it was nothing new for her to have to practice patience, and she could feel, at least, that it was just as handy for Mr. Burrage to call on the child in town, where he spent half his time, sleeping constantly at Parker's.

It was a fact that this fortunate youth called very often, and Verena saw him with Olive's full concurrence whenever she was at home. It had now been quite agreed between them that no artificial limits should be set to the famous phase; and Olive had, while it lasted, a sense of real heroism in steeling herself against uneasiness. It seemed to her, moreover, only justice that she should make some concession; if Verena made a great sacrifice of filial duty in coming to live with her (this, of course, should be permanent—she would buy off the Tarrants from year to year), she must not incur the imputation (the world would judge her, in that case, ferociously) of keeping her from forming common social ties. The friendship of a young man and a young woman was, according to the pure code of New England, a common social tie; and as the weeks elapsed Miss Chancellor saw no reason to repent of her temerity. Verena was not falling in love; she felt that she should know it, should guess it on the spot. Verena was fond of human intercourse; she was essentially a sociable creature; she liked to shine and smile and talk and listen; and so far as Henry Burrage was concerned he introduced an element of easy and convenient relaxation into a life now a good deal stiffened (Olive was perfectly willing to own it) by great civic purposes. But the girl was being saved, without interference, by the simple operation of her interest in those very designs. From this time there was no need of putting pressure on her; her own springs were working; the fire with which she glowed came from within. Sacredly, brightly single she would remain; her only espousals would be at the altar of a great cause. Olive always absented herself when Mr. Burrage was announced; and when Verena afterwards attempted to give some account of his conversation she checked her, said she would rather know nothing about it—all with a very solemn mildness; this made her feel very superior, truly noble. She knew by this time (I scarcely can tell how, since Verena could give her no report), exactly what sort of a youth Mr. Burrage was: he was weakly pretentious, softly original, cultivated eccentricity, patronized progress, liked to have mysteries, sudden appointments to keep, anonymous persons to visit, the air of leading a double life, of being devoted to a girl whom people didn't know, or at least didn't meet. Of course he liked to make an impression on Verena; but what he mainly liked was to play her off upon the other girls, the daughters of fashion, with whom he danced at Papanti's. Such were the images that proceeded from Olive's rich moral consciousness. "Well, he *is* greatly interested in our movement": so much Verena once managed to announce; but the words rather irritated Miss Chancellor, who, as we know, did not care to allow for accidental exceptions in the great masculine conspiracy.

In the month of March Verena told her that Mr. Burrage was offering matrimony—offering it with much insistence, begging that she would at least wait and think of it before giving him a final answer. Verena was evidently very glad to be able to say to Olive that she had assured him she couldn't think of it, and that if he expected this he had better not come any more. He continued to come, and it was therefore to be supposed that he had ceased to count on such a concession; it was now Olive's opinion that he really didn't desire it. She had a theory that he proposed to almost any girl who was not likely to accept him—did it because he was making a collection of such episodes—a mental album of declarations, blushes, hesitations, refusals that just missed imposing themselves as acceptances, quite as he collected enamels and Cremona violins. He would be very sorry indeed to ally himself to the house of Tarrant; but such a fear didn't prevent him from holding it becoming in a man of taste to give that encouragement to low-born girls who were pretty, for one looked out for the special cases in which, for reasons (even the lowest might have reasons), they wouldn't "rise." "I told you I wouldn't marry him, and I won't," Verena said, delightedly, to her friend; her tone suggested that a certain credit belonged to her for the way she carried out her assurance. "I never thought you would, if you didn't want to," Olive replied to this; and Verena could have no rejoinder but the good-humor that sat in her eyes, unable as she was to say that she had wanted to. They had a little discussion, however, when she intimated that she pitied him for his discomfiture, Olive's contention being that, selfish, conceited, pampered and insincere, he might properly be left now to digest his affront. Miss Chancellor felt none of the remorse now that she would have felt six months before at standing in the way of such a chance for Verena, and she would have been very angry if anyone had asked her if she were not afraid of taking too much upon herself. She would have said, moreover, that she stood in no one's way, and that even if she were not there Verena would never think seriously of a frivolous little man who fiddled while Rome was burning. This did not prevent Olive from making up her mind that they had better go to Europe in the spring; a year's residence in that quarter of the globe would be highly agreeable to Verena, and might even contribute to the evolution of her genius. It cost Miss Chancellor an effort to admit that any virtue still lingered in the elder world, and that it could have any important lesson for two such good Americans as her friend and herself; but it suited her just then to make this assumption, which was not altogether sincere. It was recommended by the idea that it would get her companion out of the way—out of the way of officious fellow-citizens—till she should be absolutely firm on her feet, and would also give greater intensity to their own long conversation. On that continent of strangers they would cleave more closely still to each other. This, of course, would be to fly before the inevitable "phase," much more than to face it; but Olive decided that if they should reach unscathed the term of their delay (the first of July) she should have faced it as much as either justice or generosity demanded. I may as well say at once that she traversed most of this period without further serious alarms and with a great many little thrills of bliss and hope.

Nothing happened to dissipate the good omens with which her partnership with Verena Tarrant was at present surrounded. They threw themselves into study; they had innumerable big books from the Athenaeum, and consumed the midnight oil. Henry Burrage, after Verena had shaken her head at him so sweetly and sadly, returned to New York, giving no sign; they only heard that he had taken refuge under the ruffled maternal wing. (Olive, at least, took for granted the wing was ruffled; she could fancy how Mrs. Burrage would be affected by the knowledge that her son had been refused by the daughter of a mesmeric healer. She would be almost as angry as if she had learnt that he had been accepted.) Matthias Pardon had not yet taken his revenge in the newspapers; he was perhaps nursing his thunderbolts; at any rate, now that the operatic season had begun, he was much occupied in interviewing the principal singers, one of whom he described in one of the leading journals (Olive, at least, was sure it was only he who could write like that), as "a dear little woman with baby dimples and kittenish movements." The Tarrants were apparently given up to a measure of sensual ease with which they had not hitherto been familiar, thanks to the increase of income that they drew from their eccentric protectress. Mrs. Tarrant now enjoyed the ministrations of a "girl"; it was partly her pride (at any rate, she chose to give it this turn), that her house had for many years been conducted without the element—so debasing on both sides—of servile, mercenary labor. She wrote to Olive (she was perpetually writing to her now, but Olive never answered), that she was conscious of having fallen to a lower plane, but she admitted that it was a prop to her wasted spirit to have someone to converse with when Selah was off. Verena, of course, perceived the difference, which was inadequately explained by the theory of a sudden increase of her father's practice (nothing of her father's had ever increased like that), and ended by guessing the cause of it—a discovery which did not in the least disturb her equanimity. She accepted the idea that her parents should receive a pecuniary tribute from the extraordinary friend whom she had encountered on the threshold of womanhood, just as she herself accepted that friend's irresistible hospitality. She had no worldly pride, no traditions of independence, no ideas of what was done and what was not done; but there was only one thing that equalled this perfectly gentle and natural insensibility to favors—namely, the inveteracy of her habit of not asking them. Olive had had an apprehension that she would flush a little at learning the terms on which they should now be able to pursue their career together; but Verena never changed color; it was either not new or not disagreeable to her that the authors of her being should be bought off, silenced by money, treated as the troublesome of the lower orders are treated when they are not locked up; so that her friend had a perception, after this, that it would probably be impossible in any way ever to offend her. She was too rancorless, too detached from conventional standards, too free from private self-reference. It was too much to say of her that she forgave injuries, since she was not conscious of them; there was in forgiveness a certain arrogance of which she was incapable, and her bright mildness glided over the many traps that life sets for our consistency. Olive had always held that pride was necessary to character, but there was no peculiarity of Verena's that could make her spirit seem less pure. The added luxuries in the little house at Cambridge, which even with their help was still such a penal settlement, made her feel afresh that before she came to the rescue the daughter of that house had traversed a desert of sordid misery. She had cooked and washed and swept and stitched; she had worked harder than any of Miss Chancellor's servants. These things had left no trace upon her person or her mind; everything fresh and fair renewed itself in her with extraordinary facility, everything ugly and tiresome evaporated as soon as it touched her; but Olive deemed that, being what she was, she had a right to immense compensations. In the future she should have exceeding luxury and ease, and Miss Chancellor had no difficulty in persuading herself that persons doing the high intellectual and moral work to which the two young ladies in Charles Street were now committed owed it to themselves, owed it to the groaning sisterhood, to cultivate the best material conditions. She herself was nothing of a sybarite, and she had proved, visiting the alleys and slums of Boston in the service of the Associated Charities, that there was no foulness of disease or misery she feared to look in the face; but her house had always been thoroughly well regulated, she was passionately clean, and she was an excellent woman of business. Now, however, she elevated daintiness to a religion; her interior shone with superfluous friction, with punctuality, with winter roses. Among these soft influences Verena herself bloomed like the flower that attains such perfection in Boston. Olive had always rated high the native refinement of her countrywomen, their latent "adaptability," their talent for accommodating themselves at a glance to changed conditions; but the way her companion rose with the level of the civilization that surrounded her, the way she assimilated all delicacies and absorbed all traditions, left this friendly theory halting behind. The winter days were still, indoors, in Charles Street, and the winter nights secure from interruption. Our two young women had plenty of duties, but Olive had never favored the custom of running in and out. Much conference on social and reformatory topics went forward under her roof, and she received her colleagues—she belonged to twenty associations and committees—only at preappointed hours, which she expected them to observe rigidly. Verena's share in these proceedings was not active; she hovered over them, smiling, listening, dropping occasionally a fanciful though never an idle word, like some gently animated image placed there for good omen. It was understood that her part was before the scenes, not behind; that she was not a prompter, but (potentially, at least) a "popular favorite," and that the work over which Miss Chancellor presided so efficiently was a general preparation of the platform on which, later, her companion would execute the most striking steps.

The western windows of Olive's drawing room, looking over the water, took in the red sunsets of winter; the long, low bridge that crawled, on its staggering posts, across the Charles; the casual patches of ice and snow; the desolate suburban horizons, peeled and made bald by the rigor of the season; the general hard, cold void of the prospect; the extrusion, at Charlestown, at Cambridge, of a few chimneys and steeples, straight, sordid tubes of factories and engineshops, or spare, heavenward finger of the New England meeting house. There was something inexorable in the poverty of the scene, shameful in the meanness of its details, which gave a collective impression of boards and tin and frozen earth, sheds and rotting piles, railwaylines striding flat across a thoroughfare of puddles, and tracks of the humbler, the universal horsecar, traversing obliquely this path of danger; loose fences, vacant lots, mounds of refuse, yards bestrewn with iron pipes, telegraph poles, and bare wooden backs of places. Verena thought such a view lovely, and she was by no means without excuse when, as the afternoon closed, the ugly picture was tinted with a clear, cold rosiness. The air, in its windless chill, seemed to tinkle like a crystal, the faintest gradations of tone were perceptible in the sky, the west became deep and delicate, everything grew doubly distinct before taking on the dimness of evening. There were pink flushes on snow, "tender" reflections in patches of stiffened marsh, sounds of car-bells, no longer vulgar, but almost silvery, on the long bridge, lonely outlines of distant dusky undulations against the fading glow. These agreeable effects used to light up that end of the drawing room, and Olive often sat at the window with her companion before it was time for the lamp. They admired the sunsets, they rejoiced in the ruddy spots projected upon the parlor wall, they followed the darkening perspective in fanciful excursions. They watched the stellar points come out at last in a colder heaven, and then, shuddering a little, arm in arm, they turned away, with a sense that the winter night was even more cruel than the tyranny of men—turned back to drawn curtains and a brighter fire and a glittering tea-tray and more and more talk about the long martyrdom of women, a subject as to which Olive was inexhaustible and really most interesting. There were some nights of deep snowfall, when Charles Street was white and muffled and the doorbell foredoomed to silence, which seemed little islands of lamplight, of enlarged and intensified vision. They read a great deal of history together, and read it ever with the same thought—that of finding confirmation in it for this idea that their sex had suffered inexpressibly, and that at any moment in the course of human affairs the state of the world would have been so much less horrible (history seemed to them in every way horrible), if women had been able to press down the scale. Verena was full of suggestions which stimulated discussions; it was she, oftenest, who kept in view the fact that a good many women in the past had been intrusted with power and had not always used it amiably, who brought up the wicked queens, the profligate mistresses of kings. These ladies were easily disposed of between the two, and the public crimes of Bloody Mary, the private misdemeanors of Faustina, wife of the pure Marcus Aurelius, were very satisfactorily classified. If the influence of women in the past accounted for every act of virtue that men had happened to achieve, it only made the matter balance properly that the influence of men should explain the casual irregularities of the other sex. Olive could see how few books had passed through Verena's hands, and how little the home of the Tarrants had been a house of reading; but the girl now traversed the fields of literature with her characteristic lightness of step. Everything she turned to or took up became an illustration of the facility, the "giftedness," which Olive, who had so little of it, never ceased to wonder at and prize. Nothing frightened her; she always smiled at it, she could do anything she tried. As she knew how to do other things, she knew how to study; she read quickly and remembered infallibly; could repeat, days afterward, passages that she appeared only to have glanced at. Olive, of course, was more and more happy to think that their cause should have the services of an organization so rare.

All this doubtless sounds rather dry, and I hasten to add that our friends were not always shut up in Miss Chancellor's strenuous parlor. In spite of Olive's desire to keep her precious inmate to herself and to bend her attention upon their common studies, in spite of her constantly reminding Verena that this winter was to be purely educative and that the platitudes of the satisfied and unregenerate would have little to teach her, in spite, in short, of the severe and constant duality of our young women, it must not be supposed that their life had not many personal confluents and tributaries. Individual and original as Miss Chancellor was universally acknowledged to be, she was yet a typical Bostonian, and as a typical Bostonian she could not fail to belong in some degree to a "set." It had been said of her that she was in it but not of it; but she was of it enough to go occasionally into other houses and to receive their occupants in her own. It was her belief that she filled her teapot with the spoon of hospitality, and made a good many select spirits feel that they were welcome under her roof at convenient hours. She had a preference for what she called *real* people, and there were several whose reality she had tested by arts known to herself. This little society was rather suburban and miscellaneous; it was prolific in ladies who trotted about, early and late, with books from the Athenæum nursed behind their muff, or little nosegays of exquisite flowers that they were carrying as presents to each other. Verena, who, when Olive was not with her, indulged in a good deal of desultory contemplation at the window, saw them pass the house in Charles Street, always apparently straining a little, as if they might be too late for something. At almost any time, for she envied their preoccupation, she would have taken the chance with them. Very often, when she described them to her mother, Mrs. Tarrant didn't know who they were; there were even days (she had so many discouragements) when it seemed as if she didn't want to know. So long as they were not someone else, it seemed to be no use that they were themselves; whoever they were, they were sure to have that defect. Even after all her mother's disquisitions Verena had but vague ideas as to whom she would have liked them to be; and it was only when the girl talked of the concerts, to all of which Olive subscribed and conducted her inseparable friend, that Mrs. Tarrant appeared to feel in any degree that her daughter was living up to the standard formed for her in their Cambridge home. As all the world knows, the opportunities in Boston for hearing good music are numerous and excellent, and it had long been Miss Chancellor's practice to cultivate the best. She went in, as the phrase is, for the superior programs, and that high, dim, dignified Music Hall, which has echoed in its time to so much eloquence and so much melody, and of which the very proportions and color seem to teach respect and attention, shed the protection of its illuminated cornice, this winter, upon no faces more intelligently upturned than those of the young women for whom Bach and Beethoven only repeated, in a myriad form, the idea that was always with them. Symphonies and fugues only stimulated their convictions, excited their revolutionary passion, led their imagination further in the direction in which it was always pressing. It lifted them to immeasurable heights; and as they sat looking at the great florid, somber organ, overhanging the bronze statue of Beethoven, they felt that this was the only temple in which the votaries of their creed could worship.

And yet their music was not their greatest joy, for they had two others which they cultivated at least as zealously. One of these was simply the society of old Miss Birdseye, of whom Olive saw more this winter than she had ever seen before. It had become apparent that her long and beautiful career was drawing to a close, her earnest, unremitting work was over, her old-fashioned weapons were broken and dull. Olive would have liked to hang them up as venerable relics of a patient fight, and this was what she seemed to do when she made the poor lady relate her battles—never glorious and brilliant, but obscure and wastefully heroic—call back the figures of her companions in arms, exhibit her medals and scars. Miss Birdseye knew that her uses were ended; she might pretend still to go about the business of unpopular causes, might fumble for papers in her immemorial satchel and think she had important appointments, might sign petitions, attend conventions, say to Doctor Prance that if she would only make her sleep she should live to see a great many improvements yet; she ached and was weary, growing almost as glad to look back (a great anomaly for Miss Birdseye) as to look forward. She let herself be coddled now by her friends of the new generation; there were days when she seemed to want nothing better than to sit by Olive's fire and ramble on about the old struggles, with a vague, comfortable sense—no physical rapture of Miss Birdseye's could be very acute—of immunity from wet feet, from the draughts that prevail at thin meetings, of independence of street-cars that would probably arrive overflowing; and also a pleased perception, not that she was an example to these fresh lives which began with more advantages than hers, but that she was in some degree an encouragement, as she helped them to measure the way the new truths had advanced—being able to tell them of such a different state of things when she was a young lady, the daughter of a very talented teacher (indeed her mother had been a teacher too), down in Connecticut. She had always had for Olive a kind of aroma of martyrdom, and her battered, unremunerated, unpensioned old age brought angry tears, springing from depths of outraged theory, into Miss Chancellor's eyes. For Verena, too, she was a picturesque humanitary figure. Verena had been in the habit of meeting martyrs from her childhood up, but she had seen none with so many reminiscences as Miss Birdseye, or who had been so nearly scorched by penal fires. She had had escapes, in the early days of abolitionism, which it was a marvel she could tell with so little implication that she had shown courage. She had roamed through certain parts of the South, carrying the Bible to the slave; and more than one of her companions, in the course of these expeditions, had been tarred and feathered. She herself, at one season, had spent a month in a Georgian jail. She had preached temperance in Irish circles where the doctrine was received with missiles; she had interfered between wives and husbands mad with drink; she had taken filthy children, picked up in the street, to her own poor rooms, and had removed their pestilent rags and washed their sore bodies with slippery little hands. In her own person she appeared to Olive and Verena a representative of suffering humanity; the pity they felt for her was part of their pity for all who were weakest and most hardly used; and it struck Miss Chancellor (more especially) that this frumpy little missionary was the last link in a tradition, and that when she should be called away the heroic age of New England life—the age of plain living and high thinking, of pure ideals and earnest effort, of moral passion and noble experiment—would effectually be closed. It was the perennial freshness of Miss Birdseye's faith that had had such a contagion for these modern maidens, the unquenched flame of her transcendentalism, the simplicity of her vision, the way in which, in spite of mistakes, deceptions, the changing fashions of reform, which make the remedies of a previous generation look as ridiculous as their bonnets, the only thing that was still actual for her was the elevation of the species by the reading of Emerson and the frequentation of Tremont Temple. Olive had been active enough, for years, in the city-missions; she too had scoured dirty children, and, in squalid lodging-houses, had gone into rooms where the domestic situation was strained and the noises made the neighbors turn pale. But she reflected that after such exertions she had the refreshment of a pretty house, a draw-ing room full of flowers, a crackling hearth, where she threw in pine-cones and made them snap, an imported tea-service, a Chickering piano, and the *Deutsche Rundschau;* whereas Miss Birdseye had only a bare, vulgar room, with a hideous flowered carpet (it looked like a dentist's), a cold furnace, the evening-paper, and Doctor Prance. Olive and Verena were present at another of her gatherings before the winter ended; it resembled the occasion that we described at the beginning of this history, with the difference that Mrs. Farrinder was not there to oppress the company with her greatness, and that Verena made a speech without the co-operation of her father. This young lady had delivered herself with even finer effect than before, and Olive could see how much she had gained, in confidence and range of allusion, since the educative process in Charles Street began. Her *motif* was now a kind of unprepared tribute to Miss Birdseye, the fruit of the occasion and of the unanimous tenderness of the younger members of the circle, which made her a willing mouthpiece. She pictured her laborious career, her early associates (Eliza P. Moseley was not ne-glected as Verena passed), her difficulties and dangers and triumphs, her humanizing effect upon so many, her serene and honored old age —expressed, in short, as one of the ladies said, just the very way they all felt about her. Verena's face brightened and grew triumphant as she spoke, but she brought tears into the eyes of most of the others. It was Olive's opinion that nothing could be more graceful and touch-ing, and she saw that the impression made was now deeper than on the former evening. Miss Birdseye went about with her eighty years of innocence, her undiscriminating spectacles, asking her friends if it wasn't perfectly splendid; she took none of it to herself, she regarded it only as a brilliant expression of Verena's gift. Olive thought, afterwards, that if a collection could only be taken up on the spot, the good lady would be made easy for the rest of her days; then she remembered that most of her guests were as impecunious as herself.

I have intimated that our young friends had a source of fortifying emotion which was distinct from the hours they spent with Beethoven and Bach, or in hearing Miss Birdseye describe Concord as it used to be. This consisted in the wonderful insight they had obtained into the history of feminine anguish. They perused that chapter perpetually and zealously, and they derived from it the purest part of their mission. Olive had pored over it so long, so earnestly, that she was now in complete possession of the subject; it was the one thing in life which she felt she had really mastered. She was able to exhibit it to Verena with the greatest authority and accuracy, to lead her up and down, in and out, through all the darkest and most tortuous passages. We know that she was without belief in her own eloquence, but she was very eloquent when she reminded Verena how the exquisite weakness of women had never been their defense, but had only exposed them to sufferings more acute than masculine grossness can conceive. Their odious partner had trampled upon them from the beginning of time, and their tenderness, their abnegation, had been his opportunity. All the bullied wives, the stricken mothers, the dishonored, deserted maidens who have lived on the earth and longed to leave it, passed and repassed before her eyes, and the interminable dim procession seemed to stretch out a myriad hands to her. She sat with them at their trembling vigils, listened for the tread, the voice, at which they grew pale and sick, walked with them by the dark waters that offered to wash away misery and shame, took with them, even, when the vision grew intense, the last shuddering leap. She had analyzed to an extraordinary fineness their susceptibility, their softness; she knew (or she thought she knew) all the possible tortures of anxiety, of suspense and dread; and she had made up her mind that it was women, in the end, who had paid for everything. In the last resort the whole burden of the human lot came upon them; it pressed upon them far more than on the others, the intolerable load of fate. It was they who sat cramped and chained to receive it; it was they who had done all the waiting and taken all the wounds. The sacrifices, the blood, the tears, the terrors were theirs. Their organism was in itself a challenge to suffering, and men had practiced upon it with an impudence that knew no bounds. As they were the weakest most had been wrung from them, and as they were the most generous they had been most deceived. Olive Chancellor would have rested her case, had it been necessary, on those general facts; and her simple and comprehensive contention was that the peculiar wretchedness which had been the very essence of the feminine lot was a monstrous artificial imposition, crying aloud for redress. She was willing to admit that women, too, could be bad; that there were many about the world who were false, immoral, vile. But their errors were as nothing to their sufferings; they had expiated, in advance, an eternity, if need be, of misconduct. Olive poured forth these views to her listening and responsive friend; she presented them again and again, and there was no light in which they did not seem to palpitate with truth. Verena was immensely wrought upon; a subtle fire passed into her; she was not so hungry for revenge as Olive, but at the last, before they went to Europe (I shall take no place to describe the manner in which she threw herself into that project), she quite agreed with her companion that after so many ages of wrong (it would also be after the European journey) men must take *their* turn, men must pay!

DOROTHY RICHARDSON
(1 8 7 3 – 1 9 5 7)

Pilgrimage

Miriam paused with her heavy bag dragging at her arm. It was a disaster. But it was the last of Mornington Road. To explain about it would be to bring Mornington Road here.

"It doesn't matter now," said Mrs. Bailey as she dropped her bag and fumbled for her purse.

"Oh, I'd better settle it at once or I shall forget about it. I'm so glad the things have come so soon."

When Mrs. Bailey had taken the half-crown they stood smiling at each other. Mrs. Bailey looked exactly as she had done the first time. It was exactly the same; there was no disappointment. The light coming through the glass above the front door made her look more shabby and worn. Her hair was more metallic. But it was the same girlish figure and the same smile triumphing over the badly fitting teeth. Miriam felt like an inmate returning after an absence. The smeariness of the marble-topped hall table did not offend her. She held herself in. It was better to begin as she meant to go on. Behind Mrs. Bailey the staircase was beckoning. There was something waiting upstairs that would be gone if she stayed talking to Mrs. Bailey.

Assuring Mrs. Bailey that she remembered the way to the room, she started at last on the journey up the many flights of stairs. The feeling of confidence that had come the first time she mounted them with Mrs. Bailey returned now. She could not remember noticing anything then but a large brown dinginess, one rich warm even tone everywhere in the house; a sharp contrast to the cold, harshly lit little bedroom in Mornington Road. The day was cold. But this house did not seem cold and, when she rounded the first flight and Mrs. Bailey was out of sight, the welcome of the place fell upon her. She knew it well, better than any place she had known in all her wanderings—the faded umbers and browns of the stair carpet, the gloomy heights of wall, a patternless sheen where the staircase lights fell upon it and, in the shadowed parts, a blurred scrolling pattern in dull madder on a brown background; the dark landings with lofty ceilings and high dark polished doors surmounted by classical reliefs in grimed plaster, the high staircase windows screened by long smoke-grimed lace curtains. On the third landing the ceiling came down nearer to the tops of the doors. The light from above made the little grained doors stare brightly. Patches of fresh brown and buff shone here and there in the threadbare linoleum. The cracks of the flooring were filled with dust and dust lay along the rim of the skirting. Two large tin trunks standing one upon the other almost barred the passage way. It was like a landing in a small suburban lodging-house, a small silent afternoon brightness, shut in and smelling of dust. Silence flooded up from the lower darkness. The hall where she had stood with Mrs. Bailey was far away below, and below that were basements deep in the earth. The outside of the house, with its first-floor balcony, the broad shallow flight of steps leading to the dark green front door, the little steep flight running sharply down into the railed area, seemed as far away as yesterday.

The little landing was a bright plateau. Under the skylight, shut off by its brightness from the rest of the house, the rooms leading from it would be bright and flat and noisy with light compared with the rest of the house. From above came the tap-tap of a door swinging gently in a breeze and behind the sound was a soft faint continuous murmur. She ran up the short twisting flight of bare stairs into a blaze of light. Would her room be a bright suburban bedroom? Had it been a dull day when she first called? The skylight was blue and gold with light, its cracks threads of bright gold. Three little glaring yellow-grained doors opened on to the small strip of uncovered dusty flooring; to the left the little box-loft, to the right the empty garret behind her own and, in front of her, her own door ajar; tapping in the breeze. The little brass knob rattled loosely in her hand and the hinge ran up the scale to a high squeak as she pushed open the door, and down again as it closed behind her neatly with a light wooden sound. The room was half dark shadow and half brilliant light.

She closed the door and stood just inside it looking at the room. It was smaller than her memory of it. When she had stood in the middle of the floor with Mrs. Bailey, she had looked at nothing but Mrs. Bailey, waiting for the moment to ask about the rent. Coming upstairs she had felt the room was hers and barely glanced at it when Mrs. Bailey opened the door. From the moment of waiting on the stone steps outside the front door, everything had opened to the movement of her impulse. She was surprised now at her familiarity with the detail of the room . . . that idea of visiting places in dreams. It was something more than that . . . all the real part of your life has a real dream in it; some of the real dream part of you coming true. You know in advance when you are really following your life. These things are familiar because reality is here. Coming events cast *light*. It is like dropping everything and walking backwards to something you know is there. However far you go out, you come back. . . . I am back now where I was before I began trying to do things like other people. I left home to get here. None of those things can touch me here.

. . . The room asserted its chilliness. But the dark yellow graining of the wall-paper was warm. It shone warmly in the stream of light pouring through the barred lattice window. In the further part of the room, darkened by the steep slope of the roof, it gleamed like stained wood. The window space was a little square wooden room, the long low double lattice breaking the roof, the ceiling and walls warmly reflecting its oblong of bright light. Close against the window was a firm little deal table covered with a thin, brightly coloured printed cotton table-cloth. When Miriam drew her eyes from its confusion of rich fresh tones, the bedroom seemed very dark. The bed, drawn in under the slope, showed an expanse of greyish white counterpane, the carpet was colourless in the gloom. She opened the door. Silence came in from the landing. The blue and gold had gone from the skylight. Its sharp grey light shone in over the dim colours of the threadbare carpet and on to the black bars of the little grate and the little strip of tarnished yellow-grained mantelpiece, running along to the bedhead where a small globeless gas bracket stuck out at an angle over the head of the bed. The sight of her luggage piled up on the other side of the fireplace drew her forward into the dimness. There was a small chest of drawers, battered and almost paintless, but with two long drawers and two small ones and a white cover on which stood a little looking-glass framed in polished pine . . . and a small yellow wardrobe with a deep drawer under the hanging part, and a little drawer in the rickety little washstand and another above the dusty cupboard of the little mahogany sideboard. I'll paint the bright part of the ceiling; scrolls of leaves. . . . Shutting the quiet door she went into the brilliance of the window space. The outside world appeared; a long row of dormer windows and the square tops of the larger windows below them, the windows black or sheeny grey in the light, cut out against the dinginess of smoke-grimed walls. The long strip of roof sloping back from the dormers was a pure even dark grey. She bent to see the sky, clear soft heavy grey, striped by the bars of her window. Behind the top rim of the iron framework of the bars was a discoloured roll of window blind. Then the bars must move. . . . Shifting the table she pressed close to the barred window. It smelt strongly of rust and dust. Outside she saw grey tiles sloping steeply from the window to a cemented gutter, beyond which was a little stone parapet about two feet high. A soft wash of madder lay along the grey tiles. There must be an afterglow somewhere, just out of sight. Her hands went through the bars and lifted the little rod which held the lattice half open. The little square four-paned frame swung free and flattened itself back against the fixed panes, out of reach, its bar sticking out over the leads. Drawing back grimed fingers and wrists striped with grime, she grasped the iron bars and pulled. The heavy framework left the window frame with a rusty creak and the sound of paint peeling and cracking. It was very heavy, but it came up and up until her arms were straight above her head, and looking up she saw a stout iron ring in a little trapdoor in the wooden ceiling and a hook in the centre of the endmost bar in the iron framework.

Kneeling on the table to raise the frame once more and fix it to the ceiling, she saw the whole length of the top row of windows across the way and wide strips of grimy stucco placed across the house fronts between the windows.

The framework of the freed window was cracked and blistered, but the little square panes were clean. There were four little windows in the row, each with four square panes. The outmost windows were immovable. The one next to the open one had lost its bar, but a push set it free and it swung wide. She leaned out, holding back from the dusty sill, and met a soft fresh breeze streaming straight in from the west. The distant murmur of traffic changed into the clear plonk plonk and rumble of swift vehicles. Right and left at the far end of the vista were glimpses of bare trees. The cheeping of birds came faintly from the distant squares and clear and sharp from neighbouring roofs. To the left the trees were black against pure grey, to the right they stood spread and bunched in front of the distant buildings blocking the vista. Running across the rose-washed façade of the central mass she could just make out "Edwards's Family Hotel" in large black letters. That was the distant view of the courtyard of Euston Station. . . . In between that and the square of trees ran the Euston Road, by day and by night, her unsleeping guardian, the rim of the world beyond which lay the northern suburbs, banished.

From a window somewhere down the street out of sight came the sound of an unaccompanied violin, clearly attacking and dropping and attacking a passage of half a dozen bars. The music stood serene and undisturbed in the air of the quiet street. The man was following the phrase, listening; strengthening and clearing it, completely undisturbed and unconscious of his surroundings. "Good heavens," she breathed quietly, feeling the extremity of relief, passing some boundary, emerging strong and equipped in a clear medium. . . . She turned back into the twilight of the room. Twenty-one and only one room to hold the richly renewed consciousness, and a living to earn, but the self that was with her in the room was the untouched tireless self of her seventeenth year and all the earlier time. The familiar light moved within the twilight, the old light. . . . She might as well wash the grime from her wrists and hands. There was a scrap of soap in the soap dish, dry and cracked and seamed with dirt. The washstand rocked as she washed her hands; the toilet things did not match, the towel-horse held one small thin face-towel and fell sideways against the wardrobe as she drew off the towel. When the gas was on she would be visible from the opposite dormer window. Short skimpy faded Madras muslin curtains screened a few inches of the endmost windows and were caught back and tied up with tape. She untied the tape and disengaged with the curtains a strong smell of dust. The curtains would cut off some of the light. She tied them firmly back and pulled at the edge of the rolled up blind. The blind, streaked and mottled with ironmould, came down in a stifling cloud of dust. She rolled it up again and washed once more. She must ask for a bath towel and do something about the blind, sponge it or something; that was all.

A light had come in the dormer on the other side of the street. It remained unscreened. Watching carefully she could see only a dim figure moving amongst motionless shapes. No need to trouble about the blind. London could come freely in day and night through the unscreened happy little panes; light and darkness and darkness and light.

London, just outside all the time, coming in with the light, coming in with the darkness, always present in the depths of the air in the room.

The gas flared out into a wide bright flame. The dingy ceiling and counterpane turned white. The room was a square of bright light and had a rich brown glow, shut brightly in by the straight square of level white ceiling and thrown up by the oblong that sloped down, white, at the side of the big bed almost to the floor. She left her things half unpacked about the floor and settled herself on the bed under the gas jet with *The Voyage of the Beagle*. Unpacking had been a distraction from the glory, very nice, getting things straight. But there was no *need* to do anything or think about anything . . . ever, here. No interruption, no one watching or speculating or treating one in some particular way that had to be met. Mrs. Bailey did not speculate. She knew, everything. Every evening here would have a glory, but not the same kind of glory. Reading would be more of a distraction than unpacking. She read a few lines. They had a fresh attractive *meaning*. Reading would be real. The dull adventures of the *Beagle* looked real, coming along through reality. She put the book on her knee and once more met the clear brown shock of her room. . . .

When she turned out the gas the window spaces remained faintly alight with a soft light like moonlight. At the window she found a soft bluish radiance cast up from below upon the opposite walls and windows. It went up into the clear blue darkness of the sky.

When she lay down the bed smelt faintly of dust. The air about her head under the sharply sloping ceiling was still a little warm with the gas. It was full of her untrammelled thoughts. Her luggage was lying about, quite near. She thought of washing in the morning in the bright light on the other side of the room . . . leaves crowding all round the lattice and here and there a pink rose . . . *several pink roses* . . . the lovely air chilling the water . . . the basin quite up against the lattice . . . dew splashing off the rose bushes in the little garden almost dark with trellises and trees, crowding with Harriett through the little damp stiff gate, the sudden lineny smell of Harriett's pinafore and the thought of Harriett in it, feeling the same, sudden bright sunshine, two shouts, great cornfields going up and up with a little track between them . . . up over Blewburton . . . *Whittenham Clumps*. Before I saw Whittenham Clumps I had always known them. But we saw them before we knew they were called Whittenham Clumps. It was a surprise to know anybody who had seen them and that they had a name.

St. Pancras bells were clamouring in the room; rapid scales, beginning at the top, coming with a loud full thump on to the fourth note and finishing with a rush to the lowest which was hardly touched before the top note hung again in the air, sounding outdoors clean and clear while all the other notes still jangled together in her room. Nothing had changed. The night was like a moment added to the day; like years going backwards to the beginning; and in the brilliant sunshine the unchanging things began again, perfectly new. She leaped out of bed into the clamorous stillness and stood in the window rolling up the warm hair that felt like a shawl round her shoulders. A cup of tea and then the bus to Harriett's. A bus somewhere just out there beyond the morning stillness of the street. What an *adventure* to go out and take a bus without having to face anybody. They were all out there, away somewhere, the very thought and sight of them, disapproving and deploring her surroundings. She listened. There they were. There were their very voices, coming plaintive and reproachful with a held-in indignation, intonations that she knew inside and out, coming on bells from somewhere beyond the squares—another church. She withdrew the coloured cover and set her spirit lamp on the inkstained table. Strong bright light was standing outside the window. The clamour of the bells had ceased. From far away down in the street a loud hoarse voice came thinly up. *"Referee—Lloyd's—Sunday Times—People—pypa. . . ."* A front door opened with a loud crackle of paint. The voice dropped to speaking tones that echoed clearly down the street and came up clear and soft and confidential. *"Referee? Lloyd's?"* The door closed with a large firm wooden sound and the harsh voice went on down the street.

St. Pancras bells burst forth again. Faintly interwoven with their bright headlong scale were the clear sweet delicate contralto of the more distant bells playing very swiftly and reproachfully a five-finger exercise in a minor key. That must be a very high-Anglican church; with light coming through painted windows on to carvings and decorations.

As she began on her solid slice of bread and butter, St. Pancras bells stopped again. In the stillness she could hear the sound of her own munching. She stared at the surface of the table that held her plate and cup. It was like sitting up to the nursery table. "How frightfully happy I am," she thought with bent head. Happiness streamed along her arms and from her head. St. Pancras bells began playing a hymn tune, in single firm beats with intervals between that left each note standing for a moment gently in the air. The first two lines were playing carefully through to the distant accompaniment of the rapid weaving and interweaving in a regular unbroken pattern of the five soft low contralto bells of the other church. The third line of the hymn ran through Miriam's head, a ding-dong to and fro from tone to semitone. The bells played it out, without the semitone, with a perfect, satisfying falsity. Miriam sat hunched against the table listening for the ascending stages of the last line. The bells climbed gently up, made a faint flat dab at the last top note, left it in the air askew above the decorous little tune and rushed away down their scale as if to cover the impropriety. They clamoured recklessly mingling with Miriam's shout of joy as they banged against the wooden walls of the window space.

Women speaking freely as poets in their own voices of their own lives is quite recent; when poor Margaret, Duchess of Newcastle, fancied herself a poet in the seventeenth century, she was driven to madness by the jeers of an uncomprehending society, and any mute, inglorious female Miltons of lesser rank were doomed to remain mute. A rash of female poets broke out in the nineteeth century, many with three names—Elizabeth Barrett Browning being the most celebrated and the model for, alas! many others—but only Emily Dickinson and Christina Rossetti were in sufficient touch with their singular lives to produce lyric poetry of permanent distinction. With the coming to birth of a new consciousness, women poets are becoming freer and their voices stronger, their talents less crippled by social conceptions of what is proper for a "female" poet (remember the days of "poetess"?). The poems of Levertov and Wakoski are the work of independent women, not needing any further introduction; such excellence speaks for itself.

DENISE LEVERTOV
(1 9 2 3 –   )

Stepping Westward

What is green in me
darkens, muscadine.

If woman is inconstant,
good, I am faithful to

ebb and flow, I fall
in season and now

is a time of ripening.
If her part

is to be true,
a north star,

good, I hold steady
in the black sky

and vanish by day,
yet burn there

in blue or above
quilts of cloud.

There is no savor
more sweet, more salt

than to be glad to be
what, woman,

and who, myself,
I am, a shadow

that grows longer as the sun
moves, drawn out

on a thread of wonder.
If I bear burdens

they begin to be remembered
as gifts, goods, a basket

of bread that hurts
my shoulders but closes me

in fragrance. I can
eat as I go.

DIANE WAKOSKI
(1 9 3 7 –   )

On Barbara's Shore

The ocean has befriended me;
the rough white crests of waves walk
as if in moccasins
trailing a pale moist shoreline.
Knowing I cannot swim,
the ocean has not taunted me; instead
its rhythms and sounds fall like sea roses
at my feet. The singer
walks on the beach/ the faces
who listen are cliffs of water. Tonight's concert
is over
but everyday of life

is some concert
clear and beautiful.
Fish, like opal roses and the singed edges
                                        of tea-garden roses
and the camelia faces of blind roses, the roe of fish,
caviar and roses,
brush past me in this life. So confusing,
but beautiful.

When I wake up in your house
the serene white rug catches my eye. I know
it reappears from room to room
a thread of continuity in all your days and nights.
I dream of living in such a well-tended house
with Eastern rugs
tiles scrubbed to Dutch freshness
copper, brass & porcelain surfaces
sustaining the day,
but in such a life
the agonies of my past would be unexplainable.
And if life is to have any ultimate agony
perhaps that pain would be
the loss of explanation/ of reason/ of meaning.

The cockatoo sits in her brass cage crushing
                                        rocks against
her powerful bird's tongue.
There is music,
poetry,
dance,
in this house.
It is clean
organized
and direct in its meanings.
My life has none of this poetry to frame it.
And being in your house
eating your generous breakfasts
sleeping in your beautiful daughter's bedroom

and talking to you
makes me feel complete
in some way
as if
each of your movements were teaching me the beauty
of women.
I know you have some wisdom
dark as a frayed ancient book
setting off your fine face
which I have not yet learned. Perhaps your
                                        greatest beauty
is a form of hope
imparted to all who talk to you,
that life will yield its secrets
to the careful searcher.
Your movements are slow
and teach me
that they can be learned.
Music of the spheres: that means the planets
                                        moving in such harmony
that beautiful chords are heard in their concert.
Surely when you move
it sets up a similar music.

All life is motion.
Mine is the quicksilver flash.
Yours
some more steady beam.
The ocean
is my background. Its motion
designing the sand,
carving the rocks,
offering life to simple things.
Your house is an ocean.
A white bird curling its feathers like flower petals
                                        against its beak.

A sea which has befriended me,
even though I cannot swim.

The Wife

"Family Happiness" is not one of Tolstoy's better-known works, but it is another example of his unrivaled ability to create women from the inside. Only Chekhov demonstrates equal understanding of such a wide range of female characters. An outwardly simple short novel of a young girl's life in the country, love for an older man, marriage to him, motherhood, disenchantment, and final acceptance of her situation, "Family Happiness" does in brief form what Tolstoy's great novels do at length—demonstrates how possibilities for moral and emotional growth come to us from the most ordinary circumstances of our life and must be grasped there and nowhere else.

I have chosen the section which includes the early days of Masha's marriage, the first quarrels, and the young wife's boredom and plea to go to St. Petersburg even though she had earlier indicated that nothing would satisfy her but a retired country life close to her husband.

It is unlikely that Tolstoy will ever be a favorite with more militant women; he treated his own wife badly and he advocated the view that women fulfilled themselves only by becoming wives and mothers (and nursing their children, which should delight the La Lèche ladies). As an artist, however, Tolstoy reached places in himself otherwise inaccessible and revealed a remarkable understanding of women. "The Kreutzer Sonata" is usually thought of as anti-female and anti-sex, a product of the more disturbed side of Tolstoy's creative power, and it certainly is an odd and unsatisfying work of art, but passages in it are both prescient and accurate, even a century later, in describing a situation that has not changed, distasteful though it may be to recognize the fact.

Two men are talking in a railroad carriage and one says:

"The education of women will always correspond to men's opinion about them. Don't we know how men regard women? . . . She is a means of enjoyment. Her body is a means of enjoyment. And she knows this. It is just as it is with slavery . . . the enslavement of woman lies simply in the fact that people desire, and think it good, to avail themselves of her as a tool of enjoyment. Well, and they liberate woman, give her all sorts of rights equal to man, but continue to regard her as an instrument of enjoyment, and so educate her in childhood and afterwards by public opinion. And there she is, still the same humiliated and depraved slave, and the man still a depraved slaveowner.

"They emancipate women in universities and in law courts, but continue to regard her as an object of enjoyment. Teach her, as she is taught among us, to regard herself as such, and she will always remain an inferior being. . . . High schools and universities cannot alter that. It can only be changed by a change in men's outlook on women and women's way of regarding themselves." *

Tolstoy's suggestions for improving this lamentable state of affairs may not be agreeable to all—virginity or lots of babies to keep Mama safe in the nursery. Still, that does not invalidate the accuracy of his analysis. At his best, when he is speaking as an artist, not as a propagandist, outraged father, guilt-ridden sensualist, or dominating husband, he remains a master of insight and understanding.

LEO TOLSTOY
(1828–1910)

Family Happiness

Days, weeks, two whole months, of seclusion in the country slipped by unnoticed, as we thought then; and yet those two months comprised feelings, emotions, and happiness, sufficient for a lifetime. Our plans for the regulation of our life in the country were not carried out at all in the way that we expected; but the reality was not inferior to our

* *The Kreutzer Sonata, The Devil and Other Tales.* Oxford World's Classics, 1950, pp. 151–152.

ideal. There was none of that hard work, performance of duty, self-sacrifice, and life for others, which I had pictured to myself before our marriage; there was, on the contrary, merely a selfish feeling of love for one another, a wish to be loved, a constant causeless gaiety and entire oblivion of all the world. It is true that my husband sometimes went to his study to work, or drove to town on business, or walked about attending to the management of the estate; but I saw what it cost him to tear himself away from me. He confessed later that every occupation, in my absence, seemed to him mere nonsense in which it was impossible to take any interest. It was just the same with me. If I read, or played the piano, or passed my time with his mother, or taught in the school, I did so only because each of these occupations was connected with him and won his approval; but whenever the thought of him was not associated with any duty, my hands fell by my sides and it seemed to me absurd to think that anything existed apart from him. Perhaps it was a wrong and selfish feeling, but it gave me happiness and lifted me high above all the world. He alone existed on earth for me, and I considered him the best and most faultless man in the world; so that I could not live for anything else than for him, and my one object was to realize his conception of me. And in his eyes I was the first and most excellent woman in the world, the possessor of all possible virtues; and I strove to be that woman in the opinion of the first and best of men.

He came to my room one day while I was praying. I looked round at him and went on with my prayers. Not wishing to interrupt me, he sat down at a table and opened a book. But I thought he was looking at me and looked round myself. He smiled, I laughed, and had to stop my prayers.

"Have you prayed already?" I asked.

"Yes. But you go on; I'll go away."

"You do say your prayers, I hope?"

He made no answer and was about to leave the room when I stopped him.

"Darling, for my sake, please repeat the prayers with me!" He stood up beside me, dropped his arms awkwardly, and began, with a serious face and some hesitation. Occasionally he turned towards me, seeking signs of approval and aid in my face.

When he came to an end, I laughed and embraced him.

"I feel just as if I were ten! And you do it all!" he said, blushing and kissing my hands.

Our house was one of those old-fashioned country houses in which several generations have passed their lives together under one roof, respecting and loving one another. It was all redolent of good sound family traditions, which as soon as I entered it seemed to become mine too. The management of the household was carried on by Tatyána Semënovna, my mother-in-law, on old-fashioned lines. Of grace and beauty there was not much; but, from the servants down to the furniture and food, there was abundance of everything, and a general cleanliness, solidity, and order, which inspired respect. The drawing-room furniture was arranged symmetrically; there were portraits on the walls, and the floor was covered with home-made carpets and mats. In the morning room there was an old piano, with chiffoniers of two different patterns, sofas, and little carved tables with bronze ornaments. My sitting room, especially arranged by Tatyána Semënovna, contained the best furniture in the house, of many styles and periods, including an old pier glass, which I was frightened to look into at first, but came to value as an old friend. Though Tatyána Semënovna's voice was never heard, the whole household went like a clock. The number of servants was far too large (they all wore soft boots with no heels, because Tatyána Semënovna had an intense dislike for stamping heels and creaking soles); but they all seemed proud of their calling, trembled before their old mistress, treated my husband and me with an affectionate air of patronage, and performed their duties, to all appearance, with extreme satisfaction. Every Saturday the floors were scoured and the carpets beaten without fail; on the first of every month there was a religious service in the house and holy water was sprinkled; on Tatyána Semënovna's name day and on her son's (and on mine too, beginning from that autumn) an entertainment was regularly provided for the whole neighborhood. And all this had gone on without a break ever since the beginning of Tatyána Semënovna's life.

My husband took no part in the household management, he attended only to the farm work and the laborers, and gave much time to this. Even in winter he got up so early that I often woke to find him gone. He generally came back for early tea, which we drank alone together; and at that time, when the worries and vexations of the farm were over, he was almost always in that state of high spirits which we called "wild ecstasy." I often made him tell me what he had been doing in the morning, and he gave such absurd accounts that we both laughed till we cried. Sometimes I insisted on a serious account, and he gave it, restraining a smile. I watched his eyes and moving lips and took nothing in: the sight of him and the sound of his voice were pleasure enough.

"Well, what have I been saying? Repeat it," he would sometimes say. But I could repeat nothing. It seemed so absurd that *he* should talk to *me* of any other subject than ourselves. As if it mattered in the least what went on in the world outside! It was at a much later time that I began to some extent to understand and take an interest in his occupations. Tatyána Semënovna never appeared before dinner: she breakfasted alone and said good-morning to us by deputy. In our exclusive little world of frantic happiness a voice from the staid orderly region in which she dwelt was quite startling: I often lost self-control and could only laugh without speaking, when the maid stood before me with folded hands and made her formal report: "The mistress bade me inquire how you slept after your walk yesterday evening; and about her I was to report that she had pain in her side all night, and a stupid dog barked in the village and kept her awake: and also I was to ask how you liked the bread this morning, and to tell you that it was not Tarás who baked today, but Nikoláshka who was trying his hand for the first time; and she says his baking is not at all bad, especially the cracknels: but the tea rusks were overbaked." Before dinner we saw little of each other: he wrote or went out again while I played the piano or read; but at four o'clock we all met in the drawing room before dinner. Tatyána Semënovna sailed out of her own room, and certain poor and pious maiden ladies, of whom there were always two or three living in the house, made their appearance also. Every day without fail my husband by old habit offered his arm to his mother, to take her in to dinner; but she insisted that I should take the other, so that every day, without fail, we stuck in the doors and got in each other's way. She also presided at dinner, where the conversation, if rather solemn, was polite and sensible. The commonplace talk between my husband and me was a pleasant interruption to the formality of those entertainments. Sometimes there were squabbles between mother and son and they bantered one another; and I especially enjoyed those scenes, because they were the best proof of the strong and tender love which united the two. After dinner Tatyána Semënovna went to the parlor, where she sat in an armchair and ground her snuff or cut the leaves of new books, while we read aloud or went off to the piano in the morning room. We read much together at this time, but music was our favorite and best enjoyment, always evoking fresh chords in our hearts and as it were revealing each afresh to the other. While I played his favorite pieces, he sat on a distant sofa where I could hardly see him. He was ashamed to betray the impression produced on him by the music; but often, when he was not expecting it, I rose from the piano, went up to him, and tried to detect on his face signs of emotion —the unnatural brightness and moistness of the eyes, which he tried in vain to conceal. Tatyána Semënovna, though she often wanted to take a look at us there, was also anxious to put no constraint upon us. So she always passed through the room with an air of indifference and a pretense of being busy; but I knew that she had no real reason for going to her room and returning so soon. In the evening I poured out tea in the large drawing room, and all the household met again. This solemn ceremony of distributing cups and glasses before the solemnly shining samovar made me nervous for a long time. I felt myself still unworthy of such a distinction, too young and frivolous to turn the tap of such a big samovar, to put glasses on Nikíta's salver, saying "For Peter Ivánovich," "For Márya Mínichna," to ask "Is it sweet enough?" and to leave out lumps of sugar for Nurse and other deserving persons. "Capital! capital! Just like a grown-up person!" was a frequent comment from my husband, which only increased my confusion.

After tea Tatyána Semënovna played patience or listened to Márya Mínichna telling fortunes by the cards. Then she kissed us both and signed us with the cross, and we went off to our own rooms. But we generally sat up together till midnight, and that was our best and pleasantest time. He told me stories of his past life; we made plans and sometimes even talked philosophy; but we tried always to speak low, for fear we should be heard upstairs and reported to Tatyána Semënovna, who insisted on our going to bed early. Sometimes we grew hungry; and then we stole off to the pantry, secured a cold supper by the good offices of Nikíta, and ate it in my sitting room by the light of one candle. He and I lived like strangers in that big old house, where the uncompromising spirit of the past and of Tatyána Semënovna ruled supreme. Not she only, but the servants, the old ladies, the furniture, even the pictures, inspired me with respect and a little alarm, and made me feel that he and I were a little out of place in that house and must always be very careful and cautious in our doings. Thinking it over now, I see that many things—the pressure of that unvarying routine, and that crowd of idle and inquisitive servants—were uncomfortable and oppressive; but at the time that very constraint made our love for one another still keener. Not I only, but he also, never grumbled openly at anything; on the contrary he shut his eyes to what was amiss. Dmítri Sídorov, one of the footmen, was a great smoker; and regularly every day, when we two were in the morning room after dinner, he went to my husband's study to take tobacco from the jar; and it was a sight to see Sergéy Mikháylych creeping on tiptoe to me with a face between delight and terror, and a wink and a warning forefinger, while he pointed at Dmítri Sídorov, who was quite unconscious of being watched. Then, when Dmítri Sídorov had gone away without having seen us, in his joy that all had passed off successfully, he declared (as he did on every other occasion) that I was a darling, and kissed me. At times his calm connivance and apparent indifference to everything annoyed me, and I took it for weakness, never noticing that I acted in the same way myself. "It's like a child who dares not show his will," I thought.

"My dear! my dear!" he said once when I told him that his weakness surprised me; "how can a man, as happy as I am, be dissatisfied with anything? Better to give way myself than to put compulsion on others; of that I have long been convinced. There is no condition in which one cannot be happy; but our life is such bliss! I simply cannot be angry; to me now nothing seems bad, but only pitiful and amusing. Above all—*le mieux est l'ennemi du bien.* Will you believe it, when I hear a ring at the bell, or receive a letter, or even wake up in the morning, I'm frightened. Life must go on, something may change; and nothing can be better than the present."

I believed him but did not understand him. I was happy; but I took that as a matter of course, the invariable experience of people in our position, and believed that there was somewhere, I knew not where, a different happiness, not greater but different.

So two months went by and winter came with its cold and snow; and, in spite of his company, I began to feel lonely, that life was repeating itself, that there was nothing new either in him or in myself,

and that we were merely going back to what had been before. He
began to give more time to business which kept him away from me,
and my old feeling returned, that there was a special department of
his mind into which he was unwilling to admit me. His unbroken
calmness provoked me. I loved him as much as ever and was as happy
as ever in his love; but my love, instead of increasing, stood still; and
another new and disquieting sensation began to creep into my heart.
To love him was not enough for me after the happiness I had felt in
falling in love. I wanted movement and not a calm course of existence.
I wanted excitement and danger and the chance to sacrifice myself for
my love. I felt in myself a superabundance of energy which found no
outlet in our quiet life. I had fits of depression which I was ashamed
of and tried to conceal from him, and fits of excessive tenderness and
high spirits which alarmed him. He realized my state of mind before
I did, and proposed a visit to St. Petersburg; but I begged him to give
this up and not to change our manner of life or spoil our happiness.
Happy indeed I was; but I was tormented by the thought that this
happiness cost me no effort and no sacrifice, though I was even painfully conscious of my power to face both. I loved him and saw that I
was all in all to him; but I wanted everyone to see our love; I wanted
to love him in spite of obstacles. My mind, and even my senses, were
fully occupied; but there was another feeling of youth and craving for
movement, which found no satisfaction in our quiet life. What made
him say that, whenever I liked, we could go to town? Had he not said
so I might have realized that my uncomfortable feelings were my own
fault and dangerous nonsense, and that the sacrifice I desired was there
before me, in the task of overcoming these feelings. I was haunted by
the thought that I could escape from depression by a mere change
from the country; and at the same time I felt ashamed and sorry to
tear him away, out of selfish motives, from all he cared for. So time
went on, the snow grew deeper, and there we remained together, all
alone and just the same as before, while outside I knew there was noise
and glitter and excitement, and hosts of people suffering or rejoicing
without one thought of us and our remote existence. I suffered most
from the feeling that custom was daily petrifying our lives into one
fixed shape, that our minds were losing their freedom and becoming
enslaved to the steady passionless course of time. The morning always
found us cheerful; we were polite at dinner, and affectionate in the

evening. "It is all right," I thought, "to do good to others and lead upright lives, as he says; but there is time for that later; and there are other things, for which the time is now or never." I wanted, not what I had got, but a life of struggle; I wanted feeling to be the guide of life, and not life to guide feeling. If only I could go with him to the edge of a precipice and say, "One step, and I shall fall over—one movement, and I shall be lost!" then, pale with fear, he would catch me in his strong arms and hold me over the edge till my blood froze, and then carry me off whither he pleased.

This state of feeling even affected my health, and I began to suffer from nerves. One morning I was worse than usual. He had come back from the estate office out of sorts, which was a rare thing with him. I noticed it at once and asked what was the matter. He would not tell me and said it was of no importance. I found out afterwards that the police inspector, out of spite against my husband, was summoning our peasants, making illegal demands on them, and using threats to them. My husband could not swallow this at once; he could not feel it merely "pitiful and amusing." He was provoked, and therefore unwilling to speak of it to me. But it seemed to me that he did not wish to speak to me about it because he considered me a mere child, incapable of understanding his concerns. I turned from him and said no more. I then told the servant to ask Márya Mínichna, who was staying in the house, to join us at breakfast. I ate my breakfast very fast and took her to the morning room, where I began to talk loudly to her about some trifle which did not interest me in the least. He walked about the room, glancing at us from time to time. This made me more and more inclined to talk and even to laugh; all that I said myself, and all that Márya Mínichna said, seemed to me laughable. Without a word to me he went off to his study and shut the door behind him. When I ceased to hear him, all my high spirits vanished at once: indeed Márya Mínichna was surprised and asked what was the matter. I sat down on a sofa without answering, and felt ready to cry. "What has he got on his mind?" I wondered; "some trifle which he thinks important; but, if he tried to tell it me, I should soon show him it was mere nonsense. But he must needs think that I won't understand, must humiliate me by his majestic composure, and always be in the right as against me. But I too am in the right when I find things tiresome and trivial," I reflected; "and I do well to want an active life rather than to stagnate in one spot and feel life flowing past me. I want to move forward, to have some new experience every day and every hour, whereas he wants to stand still and to keep me standing beside him. And how easy it would be for him to gratify me! He need not take me to town; he need only be like me and not put compulsion on himself and regulate his feelings, but live simply. That is the advice he gives me, but he is not simple himself. That is what is the matter."

I felt the tears rising and knew that I was irritated with him. My irritation frightened me, and I went to his study. He was sitting at the table, writing. Hearing my step, he looked up for a moment and then went on writing; he seemed calm and unconcerned. His look vexed me: instead of going up to him, I stood beside his writing table, opened a book, and began to look at it. He broke off his writing again and looked at me.

"Másha, are you out of sorts?" he asked.

I replied with a cold look, as much as to say, "You are very polite, but what is the use of asking?" He shook his head and smiled with a tender timid air; but his smile, for the first time, drew no answering smile from me.

"What happened to you to-day?" I asked; "why did you not tell me?"

"Nothing much—a trifling nuisance," he said. "But I might tell you now. Two of our serfs went off to the town . . ."

But I would not let him go on.

"Why would you not tell me, when I asked you at breakfast?"

"I was angry then and should have said something foolish."

"I wished to know then."

"Why?"

"Why do you suppose that I can never help you in anything?"

"Not help me!" he said, dropping his pen. "Why, I believe that without you I could not live. You not only help me in everything I do, but you do it yourself. You are very wide of the mark," he said, and laughed. "My life depends on you. I am pleased with things, only because you are there, because I need you . . ."

"Yes, I know; I am a delightful child who must be humored and kept quiet," I said in a voice that astonished him, so that he looked up as if this was a new experience; "but I don't want to be quiet and calm; that is more in your line, and too much in your line," I added.

"Well," he began quickly, interrupting me and evidently afraid to let me continue, "when I tell you the facts, I should like to know your opinion."

"I don't want to hear them now," I answered. I did want to hear the story, but I found it so pleasant to break down his composure. "I don't want to play at life," I said, "but to live, as you do yourself."

His face, which reflected every feeling so quickly and so vividly, now expressed pain and intense attention.

"I want to share your life, to . . ." but I could not go on—his face showed such deep distress. He was silent for a moment.

"But what part of my life do you not share?" he asked; "is it because I, and not you, have to bother with the inspector and with tipsy laborers?"

"That's not the only thing," I said.

"For God's sake try to understand me, my dear!" he cried. "I know that excitement is always painful; I have learned that from the experience of life. I love you, and I can't but wish to save you from excitement. My life consists of my love for you; so you should not make life impossible for me."

"You are always in the right," I said without looking at him.

I was vexed again by his calmness and coolness while I was conscious of annoyance and some feeling akin to penitence.

"Másha, what is the matter?" he asked. "The question is not, which of us is in the right—not at all; but rather, what grievance have you against me? Take time before you answer, and tell me all that is in your mind. You are dissatisfied with me: and you are, no doubt, right; but let me understand what I have done wrong."

But how could I put my feeling into words? That he understood me at once, that I again stood before him like a child, that I could do nothing without his understanding and foreseeing it—all this only increased my agitation.

"I have no complaint to make to you," I said; "I am merely bored and want not to be bored. But you say that it can't be helped, and, as always, you are right."

I looked at him as I spoke. I had gained my object: his calmness had disappeared, and I read fear and pain in his face.

"Másha," he began in a low troubled voice, "this is no mere trifle: the happiness of our lives is at stake. Please hear me out without answering. Why do you wish to torment me?"

But I interrupted him.

"Oh, I know you will turn out to be right. Words are useless; of course you are right." I spoke coldly, as if some evil spirit were speaking with my voice.

"If you only knew what you are doing!" he said, and his voice shook.

I burst out crying and felt relieved. He sat down beside me and said nothing. I felt sorry for him, ashamed of myself, and annoyed at what I had done. I avoided looking at him. I felt that any look from him at that moment must express severity or perplexity. At last I looked up and saw his eyes: they were fixed on me with a tender gentle expression that seemed to ask for pardon. I caught his hand and said,

"Forgive me! I don't know myself what I have been saying."

"But I do; and you spoke the truth."

"What do you mean?" I asked.

"That we must go to St. Petersburg," he said; "there is nothing for us to do here just now."

"As you please," I said.

He took me in his arms and kissed me.

"You must forgive me," he said; "for I am to blame."

That evening I played to him for a long time, while he walked about the room. He had a habit of muttering to himself; and when I asked him what he was muttering, he always thought for a moment and then told me exactly what it was. It was generally verse, and sometimes mere nonsense, but I could always judge of his mood by it. When I asked him now, he stood still, thought an instant, and then repeated two lines from Lérmontov:

> He in his madness prays for storms,
> And dreams that storms will bring him peace.

"He is really more than human," I thought; "he knows everything. How can one help loving him?"

I got up, took his arm, and began to walk up and down with him, trying to keep step.

"Well?" he asked, smiling and looking at me.

"All right," I whispered. And then a sudden fit of merriment came over us both: our eyes laughed, we took longer and longer steps, and rose higher and higher on tiptoe. Prancing in this manner, to the profound dissatisfaction of the butler and astonishment of my mother-in-law, who was playing patience in the parlor, we proceeded through the house till we reached the dining room; there we stopped, looked at one another, and burst out laughing.

A fortnight later, before Christmas, we were in St. Petersburg.

The remainder of this novella describes, with Tolstoy's unexcelled psychological realism, the growth of the narrator from a self-centered bride to a mature wife, mother, and mistress of a large household. Readers familiar with Tolstoy's ideas about the proper role of women —ideas he tried to realize in his life as well, with disastrous results— will recognize the force of these ideas behind the heroine's development. Our agreement with these ideas—that fecund motherhood is the crown of every woman's life, that all sexual pleasure for a woman is wicked self-indulgence—is not, however, necessary for our appreciation of the power of Tolstoy's art.

Which Chekhov story to include? "The Bride," "A Woman's Kingdom," "The Darling," "Anna Round the Neck"? Or "The Chorus Girl," "Anyuta," "Ariadne," "The Grasshopper"? Chekhov's stories and plays present a gallery of portraits—old women and young, rich and poor, women in love and in misery—that no other writer has yet been able to duplicate. One of his less familiar stories, "Big Volodya and Little Volodya," is about a kind of woman still with us, especially in our wealthier suburbs—the bored, restless, unhappily married woman who is unwilling to abandon the security of money, position, and assured admiration for an independent life. She flirts and fantasies but will never become enough of a person to break out of her unsatisfactory life—the only one, after all, for which her education and the expectations of others have prepared her.

ANTON CHEKHOV
(1860–1904)

Big Volodya and Little Volodya

"Please let me drive! I'll go and sit with the driver!" Sophia Lvovna said in a loud voice. "Wait a moment, driver! I'm coming to sit beside you!"

She stood up in the sleigh, and her husband, Vladimir Nikitich, and the friend of her childhood, Vladimir Mikhailovich, both held her hands to prevent her from falling. The troika was moving fast.

"I said she should never have touched the brandy," Vladimir Nikitich said in annoyance as he turned to his companion. "You're some fellow, eh?"

The colonel knew from experience that after even a moderate amount of drinking women like Sophia Lvovna often give way to hysterical laughter and then tears. He was afraid that when they reached home, instead of going to sleep, he would spend the night administering compresses and pouring out medicines.

"Whoa there!" Sophia Lvovna shouted. "I want to drive!"

She felt genuinely happy and on top of the world. For the last two months, ever since her wedding, she had tormented herself with the thought that she had married Colonel Yagich for his money and, as they say, *par dépit;* but that day, in a suburban restaurant, she came suddenly and finally to the conclusion that she loved him passionately. In spite of his fifty-four years he was so finely built, so agile and sinewy, and he was always making exquisite puns and accompanying gypsy bands. It is quite true that older men nowadays are a thousand times more interesting than the young: it seems as though age and youth have exchanged roles. The colonel was two years older than her father, but such a fact could have no significance when, to tell the truth, he had infinitely more vitality, vigor, and youthfulness than she had, and she was only twenty-three.

"Oh, my darling!" she thought. "How wonderful you are!"

In the restaurant she came to the conclusion that there was not one spark of her old feeling for her childhood friend left. For this friend, Vladimir Mikhailovich, or simply Volodya, she had felt only the day before an insane and desperate passion; now she was completely indifferent to him. All evening he had seemed stupid, dull, uninteresting, insignificant; and the way he coldbloodedly and continually escaped paying the restaurant checks had shocked her, and so she had only just been able to resist telling him: "Why don't you stay at home, if you are so poor?" The colonel paid for everything.

Perhaps because trees, telephone poles, and snowdrifts were flitting past her eyes, all kinds of disconnected thoughts were passing through her brain. She remembered now that the check at the restaurant amounted to a hundred and twenty rubles, and there was another hundred rubles for the gypsies, and tomorrow she could throw a thousand rubles away if she wanted to, while only two months ago, before her wedding, she had not three rubles to her name, and had to beg her father for the least little thing. How things had changed!

Her thoughts were confused. It occurred to her that when she was ten years old her present husband, Colonel Yagich, was flirting with her aunt, and everyone at home said he had ruined her, and it was perfectly true that her aunt came down to dinner with tears in her eyes and was always going off somewhere; and they said of her that she would never find any peace. He was extremely handsome in those days and had extraordinary success with women, a fact widely known in the town. They said that every day he went on a round of visits among his adorers, exactly like a doctor visiting his patients. Even now, in spite of his gray hair, wrinkles, and spectacles, his lean face, especially in profile, remained handsome.

Sophia Lvovna's father was an army doctor who had once served in the same regiment as Yagich. Volodya's father was also an army doctor; at one time he had served in the same regiment as Yagich and her father. In spite of many turbulent and complicated love affairs, Volodya had been a brilliant student, and now, having completed his course at the university with great success, he was specializing in foreign literature and, as they say, writing his dissertation. He lived in the barracks with his father, the army doctor, and although he was now thirty years old he still had no means of subsistence. As children, Sophia Lvovna and he had lived under the same roof, though in different apartments, and he often came to play with her, and they learned dancing and took French lessons together. As he grew to become a well-built, exceedingly handsome young man, she began to feel shy in his presence and fell madly in love with him, and she remained in love with him right up to the moment when she married Yagich. He, too, had been extraordinarily successful with women almost from the age of fourteen, and the women who deceived their husbands with him usually justified themselves by saying that Volodya was only a boy. Recently the story got around that when he was a student living in lodgings near the university, anyone who went to call on him would hear footsteps behind the door and there would come a whispered apology: *"Pardon, je ne suis pas seul."* Yagich was enthusiastic about him, and as Derzhavin blessed Pushkin,* so Yagich blessed the young student, solemnly regarding him as his successor; and apparently he was very fond of him. For whole hours they played billiards or piquet together without saying a word, and if Yagich drove out on his troika he always took Volodya with him; and Yagich alone was initiated into the mysteries of his dissertation. Earlier, when the colonel was younger, they were often rivals in love, but there was never any jealousy between them. In the society in which they moved, Yagich was nicknamed Big Volodya and his friend Little Volodya.

On the sleigh, besides Sophia Lvovna, Big Volodya, and Little Volodya, there was still another person—Margarita Alexandrovna, known as Rita, a cousin of Madame Yagich, a very pale woman, over thirty, with black eyebrows and wearing pince-nez; she smoked cigarettes continually even in the bitterest frosty weather: there was always cigarette ash on her knees and on the front of her dress. She spoke through her nose, drawling out each word, a coldhearted woman who could drink any amount of liqueurs and brandy without getting drunk, and she liked telling anecdotes with *double-entendres* in a tasteless way. At home she read serious magazines from morning to night, while strewing cigarette ash all over them and eating frozen apples.

"Oh, Sonya, stop behaving like a lunatic!" she said, drawling out the words. "Really, it is too silly for words!"

When they were in sight of the town gate, the troika went more

* The poet Gavril Derzhavin is said to have blessed the sixteen-year-old Pushkin in 1815.—Tr.

slowly, as houses and people began to flicker past; and now Sophia Lvovna grew quiet, nestling against her husband and surrendering to her own thoughts. Sitting opposite her was Little Volodya. Her happy, lighthearted thoughts were mingled with melancholy ones. She thought: "This man who is sitting opposite me knows I loved him, and it is very likely he believes the gossip that I married the colonel *par dépit*." Not once had she ever told him she was in love with him, and she had never wanted him to know this, and accordingly she had concealed her feelings; but from the expression on his face it was perfectly obvious that he had seen through her, and her pride suffered. The most humiliating thing was that ever since the wedding Little Volodya had been forcing his attentions upon her, and this had never happened before. He spent long hours with her in complete silence or talking about nothing at all, and even now in the sleigh, though he did not speak to her, he would gently touch her feet or her hands. It appeared that he wanted nothing more and was delighted with her marriage; it also appeared that he despised her and she excited in him an interest of a certain kind, as though she were an immoral, disreputable woman. And when her triumphant affection for her husband mingled in her soul with feelings of humiliation and wounded pride, she was overcome with a fierce resentment and wanted to sit in the coachman's box and whistle and scream at the horses.

They were just passing the nunnery when the huge sixteen-ton bell rang out. Rita crossed herself.

"Our Olga lives in the nunnery," Sophia Lvovna said, and then she crossed herself and shivered.

"Why did she enter a nunnery?" the colonel asked.

"*Par dépit*," Rita said angrily, with obvious reference to Sophia Lvovna's marriage to Yagich. "*Par dépit* is all the rage now. Defy the whole world—that's what they do. She was a furious little coquette, always giggling, and she only liked balls and cavaliers and then suddenly—she had gone away, and everyone was surprised!"

"Not true at all!" said Little Volodya, turning down the collar of his fur coat and revealing his handsome face. "It wasn't *par dépit* at all, but something quite horrible, if you please. Her brother Dmitry went to penal servitude, and no one knows where he is. Her mother died of grief." Then he turned up his collar.

"Olga did well," he added in a muffled voice. "Living as an adopted child and with that paragon of virtue Sophia Lvovna—you have to take that into account, too!"

Sophia Lvovna was well aware of the note of contempt in his voice and she wanted to say something to hurt him, but she remained silent. Once again she was overcome with a passion of remonstrance, and she rose to her feet and shouted in a tear-filled voice: "I want to go to the early service! Turn back, driver! I want to see Olga!"

They turned back, and the deep-toned nunnery bell reminded Sophia of Olga and about all Olga's life. Other church bells were also ringing. When the driver brought the troika to a stop, Sophia Lvovna jumped from the sleigh, and ran unescorted up to the gate of the nunnery.

"Please be quick!" her husband shouted after her. "We're already late!"

She went through the dark gateway and then along an avenue which led from the gateway to the largest of the churches, while the snow crackled under her feet and the church bells rang directly over her head, so that they seemed to penetrate her whole being. Then she came to the church door; there were three steps leading down, and a porch with icons on each side which smelled of incense and juniper, and then there was another door, and a dark figure opened it and bowed low to the ground. Inside the church, the service had not yet begun. One of the nuns was walking past the iconostasis and lighting the candles on the tall candlesticks, while another lit the candles on the luster. Here and there by the columns and the side chapels stood black motionless figures. "I suppose they will be standing there as they are now until tomorrow morning," Sophia Lvovna thought, and it seemed to her that everything in the church was cold, dark, and boring —more boring than a cemetery. With a bored gaze she watched those motionless figures growing colder each minute, and suddenly she felt as though a hand were squeezing her heart. She recognized Olga, who was one of the nuns, with thin shoulders, a black kerchief over her head, and quite short. She was sure she had seen her, though when Olga had entered the nunnery she was plump and seemed taller. Hesitating, completely overwhelmed by what she had seen, Sophia Lvovna went up to the nun and looked at her over her shoulder, and she was sure it was Olga.

"Olga!" she cried, and clapped her hands, and she was so tongue-tied that she could only say: "Olga!"

The nun recognized her at once, and her eyebrows rose in surprise. Both her pure, pale, freshly washed face and the white headband she wore under the wimple seemed to be shining with joy.

"God has sent a miracle!" she cried, and she clapped her thin, pale hands.

Sophia Lvovna threw her arms fiercely around her, and then kissed her. She was afraid Olga would smell the wine she had drunk.

"We were just driving past when I remembered about you," she said, breathing deeply, as though she had been hurrying. "Lord, how pale you are! I'm so glad to see you! Tell me how you are! Are you lonely here?"

Sophia Lvovna looked round at the other nuns and said softly: "There have been so many changes at home. You know I am married to Yagich—Vladimir Nikitich Yagich. I suppose you remember him. . . . I'm very happy!"

"Praise be," Olga said. "And is your father well?"

"Yes, he's well, thank you. He often asks about you. Olga, you must come and stay with us during the holidays."

"Yes, of course," Olga said, and she smiled. "I'll come the second day of the holidays."

Sophia Lvovna did not know why she began weeping. For a whole minute she wept silently, and then she dried her eyes and said: "Rita will be very sorry not to have seen you. She is here with us. Volodya's here, too. They are near the gate. How pleased they would be if you would come out and see them! Shall we go? The service hasn't begun yet."

"Yes, let's go," Olga agreed.

She crossed herself three times and went out with Sophia Lvovna to the gate.

"Are you really happy? Are you, Sophia?" she asked as they came into the open.

"Very happy!"

"Praise be!"

Big Volodya and Little Volodya jumped out of the sleigh as soon as they saw the nun, and they greeted her respectfully. They were both visibly touched by her pallor and the dark nun's costume, and they were both pleased because she remembered them and had come out to greet them. To prevent her from getting cold, Sophia wrapped her in a rug and covered her with a flap of fur coat. Sophia's tears of a few moments ago had cleansed and relieved her spirits, and she was happy now that this noisy, restless, and in fact thoroughly impure night could have such a pure and clear-cut sequel. To keep Olga a little longer by her side, she said: "Let's take her for a drive! Come in, Olga! We'll just have a short drive. . . ."

The men expected the nun to refuse—holy people do not ride around in troikas—but to their surprise she agreed and got into the sleigh. And when the sleigh was hurrying in the direction of the town gate they were all silent, while trying to keep her warm and comfortable, and they were all thinking about her past and her present. Her face was passionless, almost expressionless, cold, pale, transparent, as though water, not blood, were flowing through her veins. Only two or three years ago she had been plump and red-cheeked, and she had talked all the time about her beaux and giggled over every mortal thing.

Near the town gate the sleigh turned back, and ten minutes later they stopped outside the nunnery gate and Olga got out. Now the church bells were ringing again.

"May God be with you," Olga said, making a low bow as nuns always do.

"You'll come and visit us, won't you, Olga?"

"Yes, indeed!"

Then she left them and quickly disappeared through the dark gateway. Afterward the troika drove on again, and they were engulfed in a wave of melancholy. They were all silent. Sophia Lvovna felt as though her whole body had gone weak, and her spirits fell. It occurred to her that inviting a nun to sit in a sleigh and drive around with some drunken companions was stupid, tactless, and perhaps sacrilegious, and as her own drunkenness wore off, so she lost any desire to delude herself, and it became clear to her that she had no love for her husband and indeed could never love him, and it was all folly and stupidity. She had married him for his money, because, in the words of her school friends, he was madly rich, and because she was afraid of being an old maid like Rita, and because she was fed up with her father, the doctor, and because she wanted to annoy Little Volodya. If she could have known when she married her husband that her life would be hideous, dreadful, and burdensome, she would not have consented to the marriage for all the gold in the world. But the damage could never be undone, and she had to reconcile herself to it.

They went home. Lying in her warm soft bed and covering herself with her bedclothes, Sophia Lvovna remembered the dark doorway, the smell of incense, and the figures beside the columns, and she was terrified by the thought that these figures would remain motionless through the night, while she slept. The early service would go on forever, and would be followed by "the hours," and then by the mass, and then by the thanksgiving service. . . .

"Oh, there is a God, yes, there truly is a God, and I must surely die, and that is why sooner or later I must think about my soul, about eternal life, and about Olga. Olga is saved now—she has found the answers to all the questions about herself. . . . But what if there is no God? Then her life has come to nothing. But how has it come to nothing? Why?"

A moment later another thought entered her head: "Yes, there is a God, and death will surely come, and I must think about my soul. If Olga saw death before her this very minute, she would not be afraid. She is ready. The important thing is that she has solved the problem of life for herself. There is a God . . . yes. . . . But is there any other way out, except by entering a nunnery? Entering a nunnery means renouncing life, reducing it to zero. . . ."

Sophia Lvovna began to feel a bit frightened. She hid her head under a pillow.

"I mustn't think about it," she muttered. "No, I mustn't think about it. . . ."

Yagich was pacing the carpet in the adjoining room: there came the soft jingling sound of spurs as he surrendered to his contemplations. It occurred to Sophia Lvovna that this man was near and dear to her only because he bore the name of Vladimir: that was the only reason. She sat up in bed and called out tenderly: "Volodya!"

"What's the matter?" her husband answered.

"Nothing."

She lay down again. She heard the pealing of a bell, and perhaps it came from the same nunnery she had been visiting. Once again she remembered the dark gateway and the figures standing there, and there came to her the idea of God and of her own inevitable death, and she put her hands to her ears to keep out the sound of the bells. It occurred to her that a long, long life stretched before her until old age and death finally overcame her, and every day of her life she would have to live in close proximity to a man she did not love, this man who was now entering the bedroom and preparing to go to bed, and she would have to stifle her hopeless love for the other man, who was young and fascinating and in her eyes quite extraordinary. She looked up at her husband and tried to say good night to him, but instead she suddenly burst into tears. She was distraught.

"Well, here comes the music!" Yagich said, and he stressed the second syllable of "music."

She remained distraught until ten o'clock the next morning, when she finally stopped crying and trembling all over; her tears gave place to a terrible headache. Yagich was in a hurry to attend late mass; he was growling at the orderly who was helping him to dress in the next room. Once he came into the bedroom to fetch something, and his footsteps were attended by the soft jingling of spurs, and then he came in again wearing his epaulettes and medals, limping slightly from rheumatism, and it occurred to Sophia Lvovna that he looked and walked like a ravening beast.

She heard him ringing up someone on the telephone.

"Be so good as to connect me with the Vasilyevsky barracks," he said, and a minute later: "Vasilyevsky barracks? Would you please ask Dr. Salimovich to come to the telephone?" And then another minute later: "Who's speaking? Is that you, Volodya? Delighted. Dear boy, ask your father to come to the telephone at once. My wife is a bit upset after yesterday. Not at home, eh? Well, thank you very much. Excellent. Much obliged. *Merci*. . . ."

For the third time Yagich entered the bedroom, and he bent over the bed and made the sign of the cross over her and gave her his hand to kiss—the women who had loved him invariably kissed his hand, and he had fallen into the habit of doing this. Then, saying he would be back for dinner, he went out.

At noon the maid announced that Vladimir Mikhailovich had arrived. Though she was staggering with fatigue and a headache, Sophia Lvovna quietly slipped into her wonderful new lilac-colored dressing gown, which was trimmed with fur, and she hurriedly arranged her hair. In her heart she felt a surge of inexpressible tenderness, and she was trembling with joy and the fear that he might leave her. She wanted only one thing—to gaze upon him.

Little Volodya was properly attired for calling upon a lady: he wore a frock coat and a white tie. When Sophia Lvovna entered the drawing room he kissed her hand and genuinely offered his sympathy over her illness. When they sat down, he praised her dressing gown.

"I was absolutely shattered by the visit to Olga yesterday," she said. "At first I thought it was quite terrible, but now I envy her. She is like a rock which can never be destroyed, nothing can budge her. Tell me, Volodya, was there any other way out for her? Is burying oneself alive the answer to all life's problems? It is death, not life . . ."

Little Volodya's face was touched with deep emotion as he remembered Olga.

"Listen to me, Volodya, you are a clever man," Sophia Lvovna went on. "Teach me how to rise above myself, as she has done. Of course, I am not a believer and could never enter a nunnery, but surely I could do something which is equivalent. My life is not an easy one," she added after a pause. "Tell me something which will give me faith. Tell me something, even if it is only a single word."

"One word? Well—*ta-ra-ra-boom-dee-ay!*"

"Volodya, why do you despise me?" she asked, livid with anger. "You have a quite fatuous way of talking to me—I beg your pardon, but you do—people don't talk to their friends and women acquaintances like that. You are so successful and so learned, and you love science, yet you never talk to me about scientific things. Why? Am I not worthy?"

Little Volodya's brows were knit with vexation.

"Why this sudden interest in science?" he asked. "What about a discussion on the constitution—or maybe about sturgeon and horseradish?"

"Very well. I'm an insignificant, silly, stupid woman without principles. I have an appalling number of faults. I'm a psychopath, I am utterly depraved—I should be despised for these things. But remember, you are ten years older than I am, and my husband is thirty years older. I've grown up before your eyes, and if you had wanted, you could have made anything out of me—even an angel. But instead"—and here her voice quivered—"you treated me abominably! Yagich married me when he was already an old man, but you could have . . ."

"We've had quite enough of that, haven't we?" Volodya said, sitting close to her and kissing both her hands. "Let the Schopenhauers philosophize and prove whatever they like, while I kiss your little hands . . ."

"You despise me! If only you knew how you are making me suffer!" She spoke uncertainly, knowing already that he would not believe her. "If only you knew how much I want to change and start my life afresh! I think about it with such joy!" she went on, while tears of joy actually sprang into her eyes. "Oh, to be good, honest, pure, never to lie, to have an aim in life . . ."

"Please stop putting on those silly airs—I don't like them at all," Volodya said, and his face assumed a whimsical expression. "Dear God, it's like being on the stage! Why don't we behave like ordinary people?"

She was afraid he would be angry and go away, and so she began to justify herself, and she forced herself to smile to please him, and once again she talked about Olga and how much she wanted to solve the problem of her life and become human.

"*Ta-ra-ra-boom-dee-ay*," he sang under his breath. "*Ta-ra-ra-boom-dee-ay* . . ."

Quite suddenly he put his arm round her waist. Without knowing what she was doing she put her hands on his shoulders and for a full minute she gazed with a look of dazed rapture at his clever mocking face, his forehead, his eyes, his handsome beard.

"You have known for a long time how much I love you," she confessed to him, and she blushed painfully, and she knew her lips were twisting convulsively with shame. "I love you! Why are you torturing me?"

She closed her eyes and kissed him fiercely on the lips, and it was a full minute before she was able to put an end to the kiss, even though she knew that kissing him was improper, and that he was standing in judgment over her, and that a servant might come in at any moment.

"Oh, how you are torturing me!" she repeated.

Half an hour later, when he had got all he wanted from her, and was sitting over lunch, she knelt before him and gazed hungrily up at his face, while he told her she resembled a puppy waiting for some ham to be thrown to it. Then he sat her on one knee and danced her up and down, as though she were a child, singing: "*Ta-ra-ra-boom-dee-ay* . . . *Ta-ra-ra-boom-dee-ay* . . .*"

When he was about to leave, she asked in passionate tones: "When? Today? Where?"

She held out both arms toward his lips, as though she wanted to tear out his answer with her hands.

"Today would hardly be suitable," he told her after some thought. "Tomorrow perhaps."

And so they parted. Before dinner Sophia Lvovna went along to the nunnery to see Olga, and was told that Olga was reading the psalter over the dead somewhere. From the nunnery she went off to see her father, but he was not at home, and so she took another sleigh and drove aimlessly through the roads and side streets until evening. For some reason she kept remembering that aunt of hers whose eyes were filled with tears and who knew no peace.

That night they drove again to the restaurant outside the town in a troika and listened to the gypsies. Driving past the nunnery, Sophia Lvovna again thought about Olga, and it terrified her that for girls and women of her station in life there was no solution except to go driving around in troikas and tell lies, or else to enter a nunnery and mortify the flesh. The next day she met her lover, and afterwards she drove around the town alone with a coachman and thought about her aunt.

During the following week Little Volodya threw her over. Life went on as usual, dull, miserable, sometimes even agonizing. The colonel and Little Volodya spent long hours together at billiards or playing piquet, and Rita continued to tell her tasteless anecdotes. Sophia Lvovna wandered around in her hired sleigh and kept asking her husband to take her for a drive in a troika.

Almost every day now she went to the nunnery and bored Olga with a recital of her unbearable sufferings, and she wept and felt she was bringing something impure and pitiable and worn-out into the cell with her, while Olga, in the tone of someone mechanically repeating a lesson, told her that all this was of no importance, it would all pass away, and God would forgive her.

"HUAN, or Dispersion" is a recent story by the author of "Another View into Love" (p. 103), who has turned from poetry to prose. Although Carol Bergé clearly owes much to D. H. Lawrence, she has taken his characteristic subject—conflict between the sexes—and tough, nervous sentence, and transformed them to her different purposes. This story offers an introduction to the kind of fiction that will increasingly come from the self-discovery that women are now so barely embarked on—although Bergé's quality is rare.

CAROL BERGÉ
(1928 –   )

HUAN, or Dispersion

The man was chopping wood. It was evening, after a day or two of rain, and the air was full of sweetness. He enjoyed the scent of the cut wood, of the wet grass, the night air. They needed the wood for the fire, but he was really chopping to impress the woman who watched: she would be delighted with the line of his arm and back as he raised and slung the axe. It was true: he was strong, and his half-naked body seemed suited to the sensual pleasure of the chore. She wanted to consider him as an intellectual being, as revolving around his intelligence; this he knew. And he was out to deprive her of this image. He wanted her to want him, to want his great physical being, to want him inside her strongly. His lust for her was continuous. And she would insist on refusing him, time after time. Until he was well beyond his own mind with it all, and became more a body creature still, and opposite to that which she thought she craved. He lived inside the body which people around him recognized as his own, more and more, rather than the spirit or intelligence she would have him be. They had been married for some while now. They lived in this very house, whose fireplace devoured the wood he hewed every day or so. Usually she watched him.

"Tea is ready, why don't you come in now," she said, unable to permit him all of his physical existence. But if they were to have tea together, it would mean another conversation. And she lived from these: she seemed, these days, to care more for them than for the conversation of the bodies. He knew that this was not entirely true of her, that she was in some way lying to herself. But he could not reach her. He could not reach through her gloss, to where her own body lived. He gave one last whack to the wood and turned toward the house, toward the lit chamber where the woman waited. It was too damned bad about that woman. But he could not, would not leave off her. He had to reach for her flesh. As he came in, she was at the table, pouring. Her back was to him, the line of the nape of her neck set so that he had to touch. She recoiled from him, not a sharp gesture but more of a slight moving aside, as he had known she would. As she always did, of late. In that new way of hers.

He could not remember how it had started, this angular game they were in. They had moved to this house, maybe that was it. They had come to this house, away from most people, a month or two ago, when the season had turned. She was eager to leave the town where they had been living. It was really a small city, and the air was not clear most of the time, it was full of the smoke of the factories which supported the townspeople. But he did not make a living in the factories, although he found use for some of their products. He was a sculptor. He was a man of the mind, as far as that went: he would work on machines, yes, but of such exquisite conception that the ordinary people could not cope with even their sight; they elicited unfamiliar emotions and sensations: the effect of a hurricane, or the softness of wind just before dawn, or the sharpness of a city set into the softness of a valley. The man had a feeling for metal that was not akin to the way factory men worked with the same material. To him, the metal was alive, had its own life, was part of the earth and constantly alive, vivid. He was filled with pleasure at the possibilities of metal, and the languages open to him as he worked. He liked old metals, and often went to the town dump to find the rusted and altered pieces he liked. Their texture aroused in him a feeling he found hard to discuss. "But why do you bring this back?" she would ask him, running her finger along an edge. "I don't know, it makes sense to me, it works. It is something that has worked, has had one life, and is still working, is still alive. It is a continuation . . ." And although she too worked creating things with her hands, he thought she did not understand. She did not question him further; she did not say anything to show that she was aware.

He wondered what she would think if he told her he thought of her skin's texture with the same feeling as when he worked with the pieces of metal. He had a passion for the metal, and for her skin. He was a man whose senses were overt, available to the world. Whereas she was obverted, presenting to the world directly but in such a design as she chose and planned. And if she could manage it, she would always send him to people rather than go out to them herself. He had no feeling of resentment about this; it was part of her way, and actually he considered it attractive. She made many such demands on him. He thought of it as her "shell" and let it be. It was really all right with him, he liked to go out and talk to people, meet people; and if she didn't, then he would be glad to go. He did not consider himself to be imposed upon, since he knew all about her, her ways of being part of the world. But he did not really know her. It was that she did not like most people, and would rather not take the chances of meeting the kind she did not like and would not get along with. Not even the possibility of meeting those she might like was enough, by now. Most people offended her, with their coarse tastes, their manufactured ideas and opinions. She considered herself too fastidious for most of them and saw no need to reach out of herself. At one time she had been ready to like the people they were meeting, in a city or town where they lived, but it had changed for her. Her trust had been betrayed. And she had withdrawn from them.

Neither of them was beautiful, in the usual sense of the word. She had badly made teeth, which protruded so that they distorted her mouth, the line of her lips. But the lips were pleasant and soft, and of course her marvelous skin was so pleasing. Her eyes were myopic and set a bit too far apart, giving altogether an almost rabbity appearance to her face. Her body was well formed, round of bosom and round of hip. She took care of herself. Her clothing was well chosen. She had dark, rich hair, which they both considered most fortunate, and she took inordinate pride in it, winding the great braid of it around her head every morning and night. One of his pleasures was to watch her in this ritual, even as she would watch him as he chopped wood or took his bath. There had been, not long ago, a great sensual pleasure between them. She had dropped back from him about the time they had moved to this house.

He thought about it as they sat drinking the tea. She wanted him to talk with her and asked him about his work, about the piece on which he was now working: "Dan, how do you see that piece? Is it finished?" But he was unwilling. "Yes. I think so." But she insisted: "I mean, it seems to me it has so much similarity to the one you did last month, and that one went—" "How can you know? That was that one, and this is this." He had cut her off sharply. She stared at him. He was angry at her myopia, that permitted her, which forced him to accept the full power of her hurt look. But she was asking the wrong things of him. "The work is my idea, and I know when a thing is done. Let me be." It was one of the edges in a week of increasing tension. During the rainy days, they had been too much together, and he was full of the dreadful tension of having her body so much near to him and so unavailable to him. At night, he would reach for her, and she would withdraw, or recoil, in that quiet or subtle way she had. It was more than he could stand. He was becoming edgy and bitter. There was not so much he asked of her, he was willing to be what she wanted of him, most of the time, because it suited him; because his own nature was so close to what she herself needed of him. But their bond had been so much physical. He could not cope with her finality of refusal. If she were a cold woman; if she had never been close to him. But there was their mutual memory of passion. There had been much passion between them, until lately.

He walked away from the table, feeling her hurt. Well, then, it was no use. It was impossible that she wouldn't know what he was feeling, what was happening with them. He could not imagine where she was. She was in another place, and he would give up rather than reach for her again. In the hallway, or the little space near the doorway, he stopped, catching a glimpse of himself in the mirror she had hung. His habit was not to stop there. This time, he saw her, as she was reflected, as he had left her. She sat at the table, obverted, her face turned bluntly toward him and not seeing him, her hands limp and at ease on the table. As he watched, she moved, not knowing she was being watched. She went to the window just next to the table, and leaned softly against the frame. Her mouth moved. She reached out and touched the frame of the window, so, and then the plant that cast its leaves' weight against the shutter. And then she stopped moving. It was as if she went rigid with a thing that came from outside her, and stopped her. He himself stopped, almost stopped breathing as he saw this. It was not like her, not like anyone he had ever seen, this kind of cessation of motion. She simply stopped moving. But she was an agile woman, an active woman who rarely was still, even in relaxation. Now he could suddenly recall other times when she might have stopped in such a way, just at the edge of his sight or consciousness. It was clearly a moving over into something else. If he had not known intelligently that this particular woman was a woman with whom he had moved for some years, was in fact his woman, he might not have recognized her. She was moved to one side of herself, in a way that seemed to remove her from his sight, and even from the room in which she seemed to be standing. It was not that she had any particular tension about her, about the way that she stood. She seemed, to his eye at least, at ease. It was another quality.

He watched. And then he realized that she had stood thus for quite a while, for perhaps six or eight minutes. Her slight gesture made him realize the amount of time that had passed. He turned from the mirror and went softly out the door. Once outside, he breathed deeply. There had been a strangeness about watching; he had felt like an eaves-dropper—it was odd. It had been she who had wanted, or needed, to move out here. Was she in some way discontent with it all? It had been she who was pressured from within, who had needed to get out of that city; she who had found this house and urged him out to look at it. "I have a feeling for this place. I feel that it will be a good place for us." And he had agreed: it was a good house, solid, with more than enough room for the two of them and their equipment and interests. She could have a room for her looms, and he could use one inside room and that outbuilding for his work. There was a bit of fine land around it. The house itself was not in such good shape, but he could work at it slowly, without taking much time from his work.

They loved the house and decided to take it. It did not cost them much more than the house they left, and it was much more com-modious. The only quarrel they had concerned the use to which cer-tain rooms would be put. She had immediately claimed a southeast room for her own, wanting the light for her colors and her eyesight. But he found he had visualized using that room himself. Ordinarily, he would not argue about such a matter; he would usually give in, because it was a small thing, and it did not matter that much. He could even consider it whimsy, justifying not fighting her. But this time he had felt a pull toward that room himself, had wanted it for himself. They became set against each other. "But it cannot be so important to you," she said shrilly. "No, it must be mine. You cannot have it." He became aroused and involved. "What do you mean, I cannot have it? What right have you? What right to the room, and what right to speak in such a manner to me? I want it for the same reason you want it. I like the light in it, and the shape of the room. It suits me." "Well, but you shall not have it. I shall." "You won't get it with screaming," he said. So she moved back, and to one side, and: "Well, then, this is to be mine. There's no point in discussing it further," in a soft, ominous voice. "I'm not sure I like it, what you're doing, Della. I think we could arrive at our choices or agreements without—" "You have all that space outdoors and I must work in here most of the time. I simply have to have this room." "Very well, then. But I resent the way you've gone about this."

He had helped her move her looms and wools into the southeast room, but he could not get over being sullen and set against her, although he could not define his own need or wish for the room. She felt his mood, but she moved blindly, firmly, as if without her own will. In past places, she had had trouble always in deciding which was to be her workroom, and then in deciding what was to go where within the room. But in this room, she put things to rights at once, as if there had been a drawing on the floor, as they had on stages in plays, indicating where each object would go. He had noticed this, but had not commented upon it. It came to his mind now. From that time, she had left him, had started to leave him. "You see?" she said, in a triumph, when the moving-in was done, "this room is perfect for me. Perfect. I can work in this room." He didn't like the set of her face then. It was a new look she wore, a kind of queer triumph, which set her against him and off to herself, left him without the grace of having bent to her whim, and shut out.

So it had gone: she would go to that room and do her work, and he would go out to his outbuilding, or his tiny workroom within the house, and do his work. He would take long walks in the woods near the house. Before now, all of it had been good: they would both be im-

mersed in their work, much of the time, and come together joyously
for meals, and to make love at odd hours, and to sleep together. They
rejoiced in each other's work. He had great respect for her as an
artisan, and his confidence in her pleased her. For her part, she usually
knew what he was about in his work, and took pleasure in going with
him on his trips for materials. He had more feeling for her work than
she did for his, but that was all right, too. He knew quite well what he
was doing. He was secure in the work. She was a good wife, an excellent wife, withal. She was bright, alive, interesting, a fine lover.
Until they moved out here, and she grew abstracted, moved into
another way of being, away from him, away from their intimacy and
smoothness together.

He did not think she was behaving to spite him. She seemed to be
moving without herself, in some way quite unavailable to his understanding. She was not a malicious woman. But she would always talk
to him, would want to sit with him and talk, she would question him,
draw him out, as she had never done. She seemed pleased when they
would talk about matters that had never before concerned her: matters
of the world around them, things about business, about museums, about
the stock market, about other countries, wars, civilizations. They had
had a life pretty much removed from such realities; not by choice but
by happenstance. If either had been concerned with the rest of the
world, it had been Daniel, who would occasionally read a newspaper
or talk with other men. She had not been interested. Her limit had
been the magazines of the world of artists, mostly. When the men
would talk, she would refrain, and move aside. It was not one of her
needs, this discussion of the life outside of their lives. Until recently.
Now she would almost force him to bring her this kind of talking. She
had a new voracity for knowledge which in itself was not that unusual;
but as it was between them it made her seem a stranger to him. He
was not used to this from her. She would perch at the edge of the chair,
her eyes fixed on his face, while he would tell her whatever he knew
and respond to her questions. He would give her as much as he knew.
There was something unnatural about her at times, as if she had a
high fever. He would watch her begin to shine, or respond, as she once
had on the bed with him. But when they were in bed, there was no
talking and there was no lovemaking. He was a whole man, he was at
home in the world, he was at ease with it, he felt himself becoming

involved in some delicate kind of perversion, the natural boundaries of his life with his woman were shifted radically. It was impossible to him. It made no sense.

And now he felt like an eavesdropper, a man watching a stranger perform an act or ritual which was somehow forbidden or sacrilegious and made him uncomfortable. He was drenched with sweat. He walked down to the stream that bordered his property. It had always had a quieting effect on him, this stream: he loved to stare into the moving brown water as it passed beneath the little stone bridge that had been there for a century. There was such a kinship between the water's motion and the metals as he worked them. His metals lived in that water, lived all around him. But now all he could see and feel was the image of his wife, as she had stood at the window. And the hurt look, when he had shut her up, before that. And he was buried beneath the month that had just passed, without the closeness of their flesh. He was full of the sense of his own body, as it curved around the need for warmth and contact. He had been given the right to expect this from her. What had happened to her, who had been that close to him! As the evening light slanted upon it, the stream seemed clouded with silt from the rains. He left it. He went to his workshed. He picked up the job he had left, a wide structure of curving shapes. Although he had thought it done, it drew him now; he went further on it, changing it, further into it until it expressed his chagrin, his anger, his dissatisfaction. In an hour, it was completely changed. It was an angry piece, it had much agony in it. He was satisfied with it. Now it was done. But damned if he would tell her that, if she should ask.

When he came back to the house, it was full night. The light was on in his wife's room. He saw her outline as she moved quickly past the window. He would not go to her. He would take some tea and sit in the kitchen. He would light a fire. But above all, he would not go to her. She had not done much work in the past couple of weeks. Had begun a few things and then dropped them, or so she had reported. And then just stopped. Even as she had at the window. There had been none of the usual restlessness about her; usually, when she was not working, she would be uneven and edgy, impatiently awaiting the time when she would feel like working again, until she would be at the next tapestry or hanging or rug. Nothing would absorb her at those times. But in the last few weeks, she would simply stop. Nothing would seem to be irritating her. She had a kind of calm, or lack of motion. Then she would start at another piece, and drop it without finishing it, and start at another, and drop it as well. And would go into that sort of calm, or absence of motion. She would go about her chores without irritability, without impatience. But she would not really be there. And she would still refuse him when he reached to touch her. Without a word, she would turn from him. What did she want of him, in this new turn of her being? He knew she wanted him. They had come together like animals, wholly, mindlessly, wonderfully. He missed her, he did not want to feed her his mind any more. Yes, he had his work, and his land, and he was his own man, but he needed his woman. Her wanting him was important to him in his spirit. Although neither of them thought about it, he was slightly crippled: his right leg had been damaged in a childhood injury, so that the defect extended to his whole right side, giving him a severe limp. Between the man and his wife, this was a condition of their lives, and taken for granted. But the man had an unacknowledged need for her physical love, because of this. It was as if she accepted him as perfect. It might have been balanced out by her odd, unfortunate teeth. Yet they never discussed these things, they took each other very much for granted. They were in good health and oblivious to each other's faults. There were so many others around them who were misformed or damaged in ways far greater than they. And they were in a way of life which sought perfection in the art rather than in the person. It was a seemly arrangement.

He drank his tea and considered. He heard her moving in the room above. Their cat jumped into the curve of his arm as it rested on the table; he patted it absently. His wife had been pouring affection on the cat. When the cat was in a room where they were, it was the cat who got the attention, the affection. She would stroke it, coo to it, fondly. He wished they had a child. There would probably be no children. But she gave to the cat more than the gestures she might give to a child, because its demands upon her were far less. She would always shut it out when she did not want it around. Since they had moved here, she came from her strange withdrawals by accepting affection from the cat. And then not always; just when she pleased. There was an obstinacy about her. He was not sure what she was about. He could hear her moving above him: the steady, even rhythm of the loom, her

steps as she went to get another skein or shifted the articulations. It
went this way for perhaps half an hour: and then she came down. She
heated water and poured herself a cup. She sat down near him. Her
face was bright and fully toward him.

"How does it go?" he asked. But he was almost unwilling that she
should answer. "Oh, it's good, I want you to see it. If you want to.
I like it." He feared her. In her dreadful enthusiasms. She reached
over and patted the cat, as it lay encircled in his arm. He did want to
see what she was working on. He did not expect to learn anything new,
but this was this time and none other, and the first in a while that
she had offered to show him her work. There was always the possibility. "Yes, I would like to see." "Good. Come." She got up, touching
the cat as it rose when he did. He followed her up the stairs. The cat
scampered past them both, and he tripped on it. "Goddam that cat."
"Leave it alone." "Oh, I wouldn't harm it. You know better." But
he thought he might, by now. "Come on, Daniel. Forget it." She had
left the lights on. On the loom was a large fabric, heavy-looking, about
seven feet square. He recognized it as a blue piece which she had
started and then deserted, about the time they had first moved in. It had
been turned toward a corner of the room, half covered, since then. It
now was almost finished. It was a peculiar blue, very dull, but with a
dark radiance, as with some metals; almost a gunmetal blue, he thought.
In the center was a figure which was catlike. "A cat?" "Yes. This cat."
And she reached down and stroked their cat. "What is this, then? A
wall hanging?" He had a certain irritability in his voice. But she did
not notice. "No, it will be a blanket." "A blanket. A blanket?" "Yes.
For our room. For our bed." "No. No, it won't. Impossible." "What do
you mean? What do you mean, impossible?" For an answer, he reached
out and slapped her face sharply. She recoiled, flinched: stared at him.
"You want to call it our bed, and you make a blanket for it, and you
put a picture of the damned cat on it. Goddam you. Goddam you."
He was crying, in a fury, in a lack of understanding her, in his own
anger. And slapped her again, wildly, a gesture that had somehow a
tenderness to it, a reaching through. "With the goddam cat on it, eh?
And where are we, you and I?" And she laughed, having no other
way available. "Yes, it is the cat, and it is for our bed." It sounded to
her quite sensible; but to him it sounded like the same tone in which
she had said, "This is my room," in a kind of blind obstinacy. "Well,

my lady, if that is what you still are, *I* think *not*." And with that he reached across and swung his fist into the loom, and his other hand raked across the weaving. The wood moved and hurtled. The loom shifted and the shuttle spun loose, dragging a strand of the blue into the room, across the room, into the air between them. "Cut it loose," he said to her. "I want it." "What for?" "Just do what I say, my lady, if you know what's good for you." "But it's not done—" "Cut it loose, I said. I know what's good for you." She saw his face clearly, and she moved to cut the wool off the loom. "I can always make another," she said, nastily. But he would not answer. His position was changed. He simply waited until she had cut it loose.

When she had it free, he took it from her and spread it on the floor, in a flinging gesture. "All right," he said, quietly, firmly, "now you sit on it." "Are you crazy? Have you gone crazy? What are you doing?" "I told you to sit on it. Sit on it." He reached past her and scooped up the cat from the floor, put it out of the room, and shut the door. "Now you may sit, and be as quiet as you please. In fact, you must not move. I wonder if you can do that." "I don't know." "Yes you do. Yes, you can. You do know. You know how, quite well. Let's see how it goes right now. If you can do it at will." She stared at him in that way of hers. "You saw me. You've seen me." "I *see* you," he said. "Now let's get on with it." She sat on the blanket, the unfinished blanket toward one corner. "No, my dear. Over there. On your precious cat. In your precious room." She did not move. "No, don't begin yet. Over there, my lass." He reached and lifted her and put her down squarely on top of the woven image of the cat. "And there you shall sit, my puss, my own sweet puss, while I watch you in your ways of leaving me, and there you shall stay until we have done with all of this."

"With all of what?" she said, smiling up into his face. She knew how to be charming. She had been too charming. She had no sense of timing, however. He reached out and slapped her again. "Now shut up with that. No more of that. Talk when you are ready, but I don't want any of that intellectual smut you have been feeding yourself on. And you can bet I know the difference. Watch it." She pulled into herself and sat. The time passed. He saw her become immobile in that strange way. He watched her working backward into herself, in that way which threatened and irritated him. "All right, then," he said, breaking through it violently. "What are you about?" And she broke: she looked up and saw him; she tried to talk, and stopped, started again: "You have always taken things for granted. I never could. I don't have to now." "Now, what's that supposed to mean?" "Daniel. Everything is new for me here now." "It sounds pretty smug to me, and I still don't understand, and you'd damn well better make me understand, lady." She did seem smug. Sitting there on that blanket, which she intended to be for both of them, as if she were planning their world completely, and were working from her own design. "I wish I could tell you what I mean. It's that everything is different now. I see everything in another way. Things come to me, and if they don't, it's all right too— being out here. A lesson—I don't have to move, sometimes. It comes to me. Not as before. I want more, but I don't have to go after it. It comes to me." "What does?" "Everything. The music. The colors. The pattern—the patterns. Do you see what I mean?" "Not yet." "It's that everything is rather more simple than I had thought. So I don't have to reach. And I have room for more, and I want more." But he was still a distance from her. "Is that why you don't need me?" "I still need you. But in a different way. Another way." "Where are you? All you've wanted is talk." "Not true. But what I want is different from what you've been giving . . ." He heard her, but he was still willfully not hearing her. "It's this damned room. I'll move you out of this room." "Oh, Daniel, don't be silly!" "Don't ever say that to me again." "It doesn't matter . . ." and she smiled up at him, too quietly.

He stood over her, his foot at the edge of the blanket. "Blanket, is it. For our bed, will you." He reached down and turned her. In so doing, he stepped onto the blanket with both feet. She twisted away, as she had so often done, as if in habit, but he caught a weird little smile as she turned. He would not release her, he would not be bought again by her averting. Now it was becoming clear to him. "Not now, missy. Not *this* time." And he took her. On the blue blanket. "*This* is how it is. Yes. This. This is how it will go," he said, into her open mouth, into her ear. Pulled his head back from hers and looked into her face, into her eyes, and said, from within her, "This is where we are now." She said nothing. But his taking of her brought out her violence, her wish for violence. She did not want him to take her, not in the old way they had had between them; but now she wanted him, she wanted this. She had a spirit to match his. She struggled and he consumed her, took her with him, he moved her in a way that was not familiar to either of them. So that she became mindless again, even while struggling and moving beneath him, went along with him into a wildness beyond herself, beyond her wish to goad him or remain reserved from him. They had been away from each other for a long time, so that they came together with a strangeness and a total familiarity. He was through to her, with his will as a man. She went with him and gave over to him.

It was good. She fell back against the rug, with her head on his arm. But she wept. "Now you've spoiled it." "Now what the hell!" "I won't be able to see as clearly. The work won't go as well. I wanted to do without, and keep the energy for the rest of the life . . ." "Oh, I see," he laughed, "well, you don't have to worry. You won't lose it. You'll see." "I wanted to be clear . . ." "You aren't cut out for that. You wait and see. It's better this way, for both of us." She was crying now. In relief from tension, in disappointment at her own flesh, she cried, sobbed. He held her, and he was still a bit angry at her, but he understood, and he was amused. He was angry for the time he had been shut out, and for her reasons. Very well, then, they were her reasons, and they were her own way through to her life. If they had given her so much, he would leave them alone, although all of it seemed ridiculous to him. He knew now that she had wanted him to take her in just this way, and that she was into a real fight with himself and herself as enemies, she would have to be convinced that she would not lose her spiritual clarity. Her reasons meant her life to her, but he could modify all of it. She did not need to leave off the body-life; it had worked for her till now, it was an answer, but it was not the answer. She had not been able to say any of it to him; not knowing how he would take it, what he would do. There was no way for her to discuss it before now. He had freed her tonight.

He continued to hold her, awkwardly but firmly, with her legs still partly around him, her skirt bunched between them at their waists, her shoe indenting his calf. She was angry with her own flesh, for its betrayal of her, for wanting him as he took her. But it was no use. She was with him, in a new way, and she would have to be with him. It was over, that clarity she had found. Or maybe he was right, it would continue in another form. He had made certain that she would not want to stand in that kind of motionless removal any more. The blanket would be a blanket, as she had designed, but it would cover them in a way she had not foreseen. This pleased her. It pleased them both.

The Mother

From Jocasta to Sophie Portnoy, the Mother bulks large in literature. Why not? Only some of us are mothers; for our sins we all have
mothers. And those who lose them in childhood seem to be worse off
than the rest of us. Naturally, enough, there are many memorable
mothers in literature—Queen Gertrude, for example, Marcel's mother
and grandmother in À la Recherche du temps perdu, all those mothers
in Mauriac's fiction, the wonderful pianist mother in Rebecca West's
The Fountain Overflows (an excerpt from it regretfully omitted because of space limitations), and that extraordinary mother and stepmother from Christina Stead's The Man Who Loved Children, a book
that needs to be read complete to grasp the complexity and sweep of
Henrietta Collyer Pollitt. Based on the author's own stepmother, fierce
Henrietta battles for her life, her autonomy, and her children against
her flabby bureaucratic husband, Sam Pollitt, with the fierceness of a
hawk. Isn't she the dark side of us all at one time or another, with her
operatic laments?

> Look at me! My back's bent in two with the fruit of my womb; aren't
> you sorry to see what happened to me because of his lust?
>   I go about with a body like a football, fit to be kicked about by a
> bohunk halfback, an all-American football, because of his lust, the
> fine, pure man that won't look at women. Don't you regret my con-
> dition because of his lust? Didn't he fix me up, pin me down, make
> sure no man would look at me while he was gallivanting with his fine
> ladies?

Because of the difficulty of excerpting from this sprawling novel—
which should lay to rest once and for all the accepted belief that

women write "nice, little domestic novels"—I've not drawn from Stead's masterpiece.

The three selections which follow are short but complex and powerful. Colette's mother, the chief influence on and the great love of her life, is beautifully memorialized in her daughter's books, from which I have chosen the episodes beginning on page 227. Colette wrote about all kinds of women at all stages in their lives, her books are revelations of insight and she is a great figure in the history of women as writers and as creators of women in literature; but her pages about her mother have a special warmth.

Tillie Olsen's story is something different—here the mother herself speaks, a new voice is heard, a new piece of life is explored. Not the child, however loving, seeing only a partial figure, "Mother," but the mother herself, comes to the center in "I Stand Here Ironing," talking against the certain knowledge she has that she has failed her daughter, will be blamed for that failure, will continue to feel the sting of her own guilt, yet could do no more, was not in control of her own life, much less her daughter's, has walked in the only light available, a dim one, and has stumbled.

The story rings with the authority of experience. Olsen has underlined that authority in her essay "Silences: When Writers Don't Write," which echoes and expands the words of Tsvetaeva's letter and points to the essential similarity of woman's experiences.

. . . [what] if, in addition to the infinite capacity, to the daily responsibilities, there are children? Balzac . . . described creation in terms of motherhood. Yes, in intelligent passionate motherhood there are similarities, and in more than the toil and patience. The calling upon total capacities; the re-living and new using of the past; the comprehensions; the fascination, absorption, intensity. All almost certain death to creation. Not because the capacities to create no longer exist, or the need . . . but because the circumstances for sustained creation are almost impossible . . . More than in any human relationship, overwhelmingly more, motherhood means being instantly interruptible, responsive, responsible. Children need now . . . the very fact that these are needs of love, not duty, that one feels them as one's self; that there is no one else to be responsible for these needs, gives them primacy. It is distraction, not meditation, that becomes habitual; interruption, not continuity; spasmodic, not constant toil.

Out of that knowledge, out of her long-delayed and nurtured art, she has written this story and others, breaking new soil.

TILLIE OLSEN
(1913–)

I Stand Here Ironing

I stand here ironing, and what you asked me moves tormented back and forth with the iron.

"I wish you would manage the time to come in and talk with me about your daughter. I'm sure you can help me understand her. She's a youngster who needs help and whom I'm deeply interested in helping."

"Who needs help?" Even if I came what good would it do? You think because I am her mother I have a key, or that in some way you could use me as a key? She has lived for nineteen years. There is all that life that has happened outside of me, beyond me.

And when is there time to remember, to sift, to weigh, to estimate, to total? I will start and there will be an interruption and I will have to gather it all together again. Or I will become engulfed with all I did or did not do, with what should have been and what cannot be helped.

She was a beautiful baby. The first and only one of our five that was beautiful at birth. You do not guess how new and uneasy her tenancy in her now-loveliness. You did not know her all those years she was thought homely, or see her poring over her baby pictures, making me tell her over and over how beautiful she had been—and would be, I would tell her—and was now, to the seeing eye. But the seeing eyes were few or nonexistent. Including mine.

I nursed her. They feel that's important nowadays. I nursed all the children, but with her, with all the fierce rigidity of first motherhood, I did like the books said. Though her cries battered me to trembling and my breasts ached with swollenness, I waited till the clock decreed.

Why do I put that first? I do not even know if it matters, or if it explains anything.

She was a beautiful baby. She blew shining bubbles of sound. She loved motion, loved light, loved color and music and textures. She would lie on the floor in her blue overalls patting the surface so hard in ecstasy her hands and feet would blur. She was a miracle to me, but when she was eight months old I had to leave her daytimes with the woman downstairs to whom she was no miracle at all, for I worked or looked for work and for Emily's father, who "could no longer endure" (he wrote in his goodbye note) "sharing want" with us.

I was nineteen. It was the pre-relief, pre-WPA world of the depression. I would start running as soon as I got off the streetcar, running up the stairs, the place smelling sour, and, awake or asleep to startle awake, when she saw me she would break into a clogged weeping that could not be comforted, a weeping I can yet hear.

After a while I found a job hashing at night so I could be with her days, and it was better. But it came to where I had to bring her to his family and leave her.

It took a long time to raise the money for her fare back. Then she got chicken pox and I had to wait longer. When she finally came, I hardly knew her, walking quick and nervous like her father, looking like her father, thin, and dressed in a shoddy red that yellowed her skin and glared at the pock marks. All the baby loveliness gone.

She was two. Old enough for nursery school, they said, and I did not know then what I know now—the fatigue of the long day, and the lacerations of group life in the nurseries that are only parking places for children.

Except that it would have made no difference if I had known. It was the only place there was. It was the only way we could be together, the only way I could hold a job.

And even without knowing, I knew. I knew that the teacher was evil because all these years it has curdled into my memory, the little boy hunched in the corner, her rasp, "Why aren't you outside, because Alvin hits you? That's no reason, go out, coward." I knew Emily hated it even if she did not clutch and implore, "Don't go, Mommy," like the other children, mornings.

She always had a reason why we should stay home. Momma, you look sick, Momma. I feel sick. Momma, the teachers aren't there today, they're sick. Momma there was a fire there last night. Momma it's a holiday today, no school, they told me.

But never a direct protest, never rebellion. I think of our others in their three-, four-year-oldness—the explosions, the tempers, the denunciations, the demands—and I feel suddenly ill. I stop the ironing. What in me demanded that goodness in her? And what was the cost, the cost to her of such goodness?

The old man living in the back once said in his gentle way, "You should smile at Emily more when you look at her." What *was* in my face when I looked at her? I loved her. There were all the acts of love.

It was only with the others I remembered what he said, so that it was the face of joy, and not of care or tightness or worry I turned to them—but never to Emily. She does not smile easily, let alone almost always, as her brothers and sisters do. Her face is closed and somber, but when she wants, how fluid. You must have seen it in her pantomimes, you spoke of her rare gift for comedy on the stage that rouses a laughter out of the audience so dear they applaud and applaud and do not want to let her go.

Where does it come from, that comedy? There was none of it in her when she came back to me that second time, after I had had to send her away again. She had a new daddy now to learn to love, and I think perhaps it was a better time. Except when we left her alone nights, telling ourselves she was old enough.

"Can't you go some other time, Mommy, like tomorrow?" she would ask. "Will it be just a little while you'll be gone?"

The time we came back, the front door open, the clock on the floor in the hall. She rigid awake. "It wasn't just a little while. I didn't cry. I called you a little, just three times, and then I went downstairs to open the door so you could come faster. The clock talked loud, I threw it away, it scared me what it talked."

She said the clock talked loud that night I went to the hospital to have Susan. She was delirious with the fever that comes before red measles, but she was fully conscious all the week I was gone and the week after we were home when she could not come near the baby or me.

She did not get well. She stayed skeleton thin, not wanting to eat, and night after night she had nightmares. She would call for me, and I would sleepily call back, "You're all right, darling, go to sleep, it's just a dream," and if she still called, in a sterner voice, "Now go to sleep, Emily, there's nothing to hurt you." Twice, only twice, when I had to get up for Susan anyhow, I went in to sit with her.

Now when it is too late (as if she would let me hold and comfort her like I do the others) I get up and go to her at her moan or restless stirring. "Are you awake? Can I get you something?" And the answer is always the same: "No, I'm all right, go back to sleep, Mother."

They persuaded me at the clinic to send her away to a convalescent home in the country where "she can have the kind of food and care you can't manage for her, and you'll be free to concentrate on the new baby." They still send children to that place. I see pictures on the society page of sleek young women planning affairs to raise money for it, or dancing at the affairs, or decorating Easter eggs or filling Christmas stockings for the children.

They never have a picture of the children, so I do not know if they still wear those gigantic red bows and the ravaged looks on the every other Sunday when parents can come to visit "unless otherwise notified"—as we were notified the first six weeks.

Oh, it is a handsome place, green lawns and tall trees and fluted flower beds. High up on the balconies of each cottage the children stand, the girls in their red bows and white dresses, the boys in white suits and giant red ties. The parents stand below shrieking up to be heard and the children shriek down to be heard, and between them the invisible wall "Not To Be Contaminated by Parental Germs or Physical Affection."

There was a tiny girl who always stood hand in hand with Emily. Her parents never came. One visit she was gone. "They moved her to Rose Cottage," Emily shouted in explanation. "They don't like you to love anybody here."

She wrote once a week, the labored writing of a seven-year-old. "I am fine. How is the baby. If I write my leter nicly I will have a star. Love." There never was a star. We wrote every other day, letters she could never hold or keep but only hear read—once. "We simply do not have room for children to keep any personal possessions," they patiently explained when we pieced one Sunday's shrieking together to plead how much it would mean to Emily to keep her letters and cards.

Each visit she looked frailer. "She isn't eating," they told us. (They had runny eggs for breakfast or mush with lumps, Emily said later, I'd hold it in my mouth and not swallow. Nothing ever tasted good, just when they had chicken.)

It took us eight months to get her released home, and only the fact that she gained back so little of her seven lost pounds convinced the social worker.

I used to try to hold and love her after she came back, but her body would stay stiff, and after a while she'd push away. She ate little. Food sickened her, and I think much of life too. Oh, she had physical lightness and brightness, twinkling by on skates, bouncing like a ball up and down up and down over the jump rope, skimming over the hill; but these were momentary.

She fretted about her appearance, thin and dark and foreign-looking at a time when every little girl was supposed to look or thought she should look a chubby blond replica of Shirley Temple. The doorbell sometimes rang for her, but no one seemed to come and play in the house or be a best friend. Maybe because we moved so much.

There was a boy she loved painfully through two school semesters. Months later she told me how she had taken pennies from my purse to buy him candy. "Licorice was his favorite and I brought him some every day, but he still liked Jennifer better'n me. Why Mommy why?" A question I could never answer.

School was a worry to her. She was not glib or quick, in a world where glibness and quickness were easily confused with ability to learn. To her overworked and exasperated teachers she was an over-conscientious "slow learner" who kept trying to catch up and was absent entirely too often.

I let her be absent, though sometimes the illness was imaginary. How different from my now-strictness about attendance with the others. I wasn't working. We had a new baby, I was home anyhow. Sometimes, after Susan grew old enough, I would keep her home from school, too, to have them all together.

Mostly Emily had asthma, and her breathing, harsh and labored, would fill the house with a curiously tranquil sound. I would bring the two old dresser mirrors and her boxes of collections to her bed. She would select beads and single earrings, bottle tops and shells, dried flowers and pebbles, old postcards and scraps, all sorts of oddments; then she and Susan would play Kingdom, setting up landscapes and furniture, peopling them with action.

Those were the only times of peaceful companionship between her and Susan. I have edged away from it, that poisonous feeling between them, that terrible balancing of hurts and needs I had to do between the two, and did so badly, those earlier years.

Oh, there are conflicts between the others too, each one human, needing, demanding, hurting, taking—but only between Emily and Susan, no, Emily toward Susan, that corroding resentment. It seems so obvious on the surface, yet it is not obvious. Susan, the second child, Susan, golden and curly-haired and chubby, quick and articulate and assured, everything in appearance and manner Emily was not; Susan, not able to resist Emily's precious things, losing or sometimes clumsily breaking them; Susan telling jokes and riddles to company for applause while Emily sat silent (to say to me later: That was *my* riddle, Mother, I told it to Susan); Susan, who for all the five years' difference in age was just a year behind Emily in developing physically.

I am glad for that slow physical development that widened the difference between her and her contemporaries, though she suffered over it. She was too vulnerable for that terrible world of youthful competition, of preening and parading, of constant measuring of yourself against every other, of envy, "If I had that copper hair," or "If I had that skin . . ." She tormented herself enough about not looking like the others, there was enough of the unsureness, the having to be conscious of words before you speak, the constant caring—what are they thinking of me? what kind of an impression am I making—there was enough without having it all magnified unendurably by the merciless physical drives.

Ronnie is calling. He is wet and I change him. It is rare there is such a cry now. That time of motherhood is almost behind me when the ear is not one's own but must always be racked and listening for the child cry, the child call. We sit for a while and I hold him, looking out over the city spread in charcoal with its soft aisles of light. "Shuggily," he breathes. A funny word, a family word, inherited from Emily, invented by her to say comfort.

In this and other ways she leaves her seal, I say aloud. And startle at my saying it. What do I mean? What did I start to gather together, to try and make coherent? I was at the terrible, growing years. War years. I do not remember them well. I was working, there were four smaller ones now, there was not time for her. She had to help be a mother, and housekeeper, and shopper. She had to set her seal. Mornings of crisis and near-hysteria trying to get lunches packed, hair combed, coats and shoes found, everyone to school or Child Care on time, the baby ready for transportation. And always the paper scribbled on by a smaller one, the book looked at by Susan, then mislaid, the homework not done. Running out to that huge school where she was one, she was lost, she was a drop; suffering over the unpreparedness, stammering and unsure in her classes.

There was so little time left at night after the kids were bedded down. She would struggle over books, always eating (it was in those years she developed her enormous appetite that is legendary in our family) and I would be ironing, or preparing food for the next day, or writing V-mail to Bill, or tending the baby. Sometimes, to make me laugh, or out of her despair, she would imitate happenings or types at school.

I think I said once, "Why don't you do something like this in the school amateur show?" One morning she phoned me at work, hardly understandable through the weeping: "Mother, I did it. I won, I won; they gave me first prize; they clapped and clapped and wouldn't let me go."

Now suddenly she was Somebody, and as imprisoned in her difference as in anonymity.

She began to be asked to perform at other high schools, even in colleges, then at city and state-wide affairs. The first one we went to, I only recognized her that first moment when, thin, shy, she almost drowned herself into the curtains. Then: Was this Emily? the control, the command, the convulsing and deadly clowning, the spell, then the roaring, stamping audience, unwilling to let this rare and precious laughter out of their lives.

Afterwards: You ought to do something about her with a gift like that—but without money or knowing how, what does one do? We have left it all to her, and the gift has as often eddied inside, clogged and clotted, as been used and growing.

She is coming. She runs up the stairs two at a time with her light graceful step, and I know she is happy tonight. Whatever it was that occasioned your call did not happen today.

"Aren't you ever going to finish the ironing, Mother? Whistler painted his mother in a rocker. I'd have to paint mine standing over an ironing board." This is one of her communicative nights and she tells me everything and nothing as she fixes herself a plate of food out of the icebox.

She is so lovely. Why did you want me to come in at all? Why were you concerned? She will find her way.

She starts up the stairs to bed. "Don't get me up with the rest in the morning." "But I thought you were having midterms." "Oh, those," she comes back in and says quite lightly, "in a couple of years when we'll all be atom-dead they won't matter a bit."

She has said it before. She believes it. But because I have been dredging the past, and all that compounds a human being is so heavy and meaningful in me, I cannot endure it tonight.

I will never total it all now. I will never come in to say: She was a child seldom smiled at. Her father left me before she was a year old. I worked her first six years when there was work, or I sent her home and to his relatives. There were years she had care she hated. She was dark and thin and foreign-looking in a world where the prestige went to blondness and curly hair and dimples, slow where glibness was prized. She was a child of anxious, not proud, love. We were poor and could not afford for her the soil of easy growth. I was a young mother, I was a distracted mother. There were the other children pushing up, demanding. Her younger sister was all that she was not. She did not like me to touch her. She kept too much in herself, her life was such she had to keep too much in herself. My wisdom came too late. She has much in her and probably nothing will come of it. She is a child of her age, of depression, of war, of fear.

Let her be. So all that is in her will not bloom—but in how many does it? There is still enough left to live by. Only help her to believe—help make it so there is cause for her to believe that she is more than this dress on the ironing board, helpless before the iron.

SIDONIE GABRIELLE COLETTE
(1873–1954)

My Mother's House:
Where Are the Children?

The house was large, topped by a lofty garret. The steep gradient of the street compelled the coach houses, stables, and poultry house, the laundry and the dairy, to huddle on a lower level all around a closed courtyard.

By leaning over the garden wall, I could scratch with my finger the poultry-house roof. The upper garden overlooked the lower garden—a warm, confined enclosure reserved for the cultivation of aubergines and pimentos—where the smell of tomato leaves mingled in July with that of the apricots ripening on the walls. In the upper garden were two twin firs, a walnut tree whose intolerant shade killed any flowers beneath it, some rosebushes, a neglected lawn and a dilapidated arbor. At the bottom, along the Rue des Vignes, a boundary wall reinforced with a strong iron railing ought to have ensured the privacy of the two gardens, but I never knew those railings other than twisted and torn from their cement foundations, and grappling in mid-air with the invincible arms of a hundred-year-old wisteria.

In the Rue de l'Hospice, a two-way flight of steps led up to the front door in the gloomy façade with its large bare windows. It was the typical burgher's house in an old village, but its dignity was upset a little by the steep gradient of the street, the stone steps being lopsided, ten on one side and six on the other.

A large solemn house, rather forbidding, with its shrill bell and its carriage entrance with a huge bolt like an ancient dungeon, a house that smiled only on its garden side. The back, invisible to passersby, was a sun-trap, swathed in a mantle of wisteria and bignonia too heavy for the trellis of worn ironwork, which sagged in the middle like a hammock and provided shade for the little flagged terrace and the threshold of the sitting room.

Is it worthwhile, I wonder, seeking for adequate words to describe the rest? I shall never be able to conjure up the splendor that adorns, in my memory, the ruddy festoons of an autumn vine borne down by its own weight and clinging despairingly to some branch of the fir trees. And the massive lilacs, whose compact flowers—blue in the shade and purple in the sunshine—withered so soon, stifled by their own exuberance. The lilacs long since dead will not be revived at my bidding, any more than the terrifying moonlight—silver, quicksilver, leaden-gray, with facets of dazzling amethyst or scintillating points of sapphire—all depending on a certain pane in the blue glass window of the summerhouse at the bottom of the garden.

Both house and garden are living still, I know; but what of that, if the magic has deserted them? If the secret is lost that opened to me a whole world—light, scents, birds and trees in perfect harmony, the murmur of human voices now silent forever—a world of which I have ceased to be worthy?

It would happen sometimes long ago, when this house and garden harbored a family, that a book lying open on the flagstones of the terrace or on the grass, a skipping rope twisted like a snake across the path, or perhaps a miniature garden, pebble-edged and planted with decapitated flowers, revealed both the presence of children and their varying ages. But such evidence was hardly ever accompanied by childish shouts or laughter, and my home, though warm and full, bore an odd resemblance to those houses which, once the holidays have come to an end, are suddenly emptied of joy. The silence, the muted breeze of the enclosed garden, the pages of the book stirred only by invisible fingers, all seemed to be asking, "Where are the children?"

It was then, from beneath the ancient iron trellis sagging to the left under the wisteria, that my mother would make her appearance, small and plump in those days when age had not yet wasted her. She would scan the thick green clumps and, raising her head, fling her call into the air: "Children! Where are the children?"

Where indeed? Nowhere. My mother's cry would ring through the garden, striking the great wall of the barn and returning to her as a faint exhausted echo. "Where . . . ? Children . . . ?"

Nowhere. My mother would throw back her head and gaze heavenward, as though waiting for a flock of winged children to alight from the skies. After a moment she would repeat her call; then, grown tired of questioning the heavens, she would crack a dry poppy head with her fingernail, rub the greenfly from a rose shoot, fill her pockets with unripe walnuts, and return to the house shaking her head over the vanished children.

And all the while, from among the leaves of the walnut tree above her gleamed the pale, pointed face of a child who lay stretched like a tomcat along a big branch, and never uttered a word. A less short-sighted mother might well have suspected that the spasmodic salutations exchanged by the twin tops of the two firs were due to some influence other than that of the sudden October squalls! And in the square dormer, above the pulley for hauling up fodder, would she not have perceived, if she had screwed up her eyes, two pale patches among the hay—the face of a young boy and the pages of his book?

But she had given up looking for us, had despaired of trying to reach us. Our uncanny turbulence was never accompanied by any sound. I do not believe there can ever have been children so active and so mute. Looking back at what we were, I am amazed. No one had imposed upon us either our cheerful silence or our limited sociability. My nineteen-year-old brother, engrossed in constructing some hydro-therapeutic apparatus out of linen bladders, strands of wire and glass tubes, never prevented the younger, aged fourteen, from disemboweling a watch or from transposing on the piano, with never a false note, a melody or an air from a symphony heard at a concert in the county town. He did not even interfere with his junior's incomprehensible passion for decorating the garden with little tombstones cut out of cardboard, and each inscribed, beneath the sign of the cross, with the names, epitaph, and genealogy of the imaginary person deceased.

My sister with the too long hair might read forever with never a pause; the two boys would brush past her as though they did not see the young girl sitting abstracted and entranced, and never bother her. When I was small, I was at liberty to keep up as best I could with my long-legged brothers as they ranged the woods in pursuit of swallow-tails, white admirals, purple emperors, or hunted for grass snakes, or gathered armfuls of the tall July foxgloves which grew in the clearings already aglow with patches of purple heather. But I followed them in silence, picking blackberries, bird cherries, a chance wildflower, or roving the hedgerows and waterlogged meadows like an independent dog out hunting on its own.

"Where are the children?" She would suddenly appear like an over-solicitous mother bitch breathlessly pursuing her constant quest, head lifted and scenting the breeze. Sometimes her white linen sleeves bore witness that she had come from kneading dough for cakes or making the pudding that had a velvety hot sauce of rum and jam. If she had been washing the Havanese bitch, she would be enveloped in a long blue apron, and sometimes she would be waving a banner of rustling yellow paper, the paper used around the butcher's meat, which meant that she hoped to reassemble, at the same time as her elusive children, her carnivorous family of vagabond cats.

To her traditional cry she would add, in the same anxious and appealing key, a reminder of the time of day. "Four o'clock, and they haven't come in to tea! Where are the children? . . ." "Half-past six! Will they come home to dinner? Where are the children? . . ." That lovely voice; how I should weep for joy if I could hear it now! Our only sin, our single misdeed, was silence, and a kind of miraculous vanishing. For perfectly innocent reasons, for the sake of a liberty that no one denied us, we clambered over the railing, leaving behind our shoes, and returned by way of an unnecessary ladder or a neighbor's low wall.

Our anxious mother's keen sense of smell would discover on us traces of wild garlic from a distant ravine or of marsh mint from a treacherous bog. The dripping pocket of one of the boys would disgorge the bathing slip worn in malarial ponds, and the "little one," cut about the knees and skinned at the elbows, would be bleeding complacently under plasters of cobweb and wild pepper bound on with rushes.

"Tomorrow I shall keep you locked up! All of you, do you hear, every single one of you!"

Tomorrow! Next day the eldest, slipping on the slated roof where he was fitting a tank, broke his collarbone and remained at the foot of the wall waiting, politely silent and half unconscious, until someone came to pick him up. Next day an eighteen-rung ladder crashed plumb on the forehead of the younger son, who never uttered a cry, but brought home with becoming modesty a lump like a purple egg between his eyes.

"Where are the children?"

Two are at rest. The others grow older day by day. If there be a place of waiting after this life, then surely she who so often waited for us has not ceased to tremble for those two who are yet alive.

For the eldest of us all, at any rate, she has done with looking at the dark windowpane every evening and saying, "I feel that child is not happy. I feel she is suffering." And for the elder of the boys she no longer listens, breathlessly, to the wheels of a doctor's trap coming over the snow at night, or to the hoofbeats of the gray mare.

But I know that for the two who remain she seeks and wanders still, invisible, tormented by her inability to watch over them enough.

"Where, oh where are the children? . . ."

The Savage

She was eighteen years old when, in about 1853, he carried her off from her family, consisting of two brothers only, French journalists married and settled in Belgium, and from her friends—painters, musicians and poets—an entire Bohemia of young French and Belgian artists. A fair-haired girl, not particularly pretty, but attractive, with a wide mouth, a pointed chin and humorous gray eyes, and her hair gathered into a precarious knot slipping from its hairpins at the nape of her neck. An emancipated girl, accustomed to the frank companionship of boys, her brothers and their friends. A dowerless young woman, without trousseau or jewels, but with a slender supple body above her voluminous skirts: a young woman with a neat waist and softly rounded shoulders, small and sturdy.

The Savage saw her on a summer's day while she was spending a few weeks with her peasant foster-mother on a visit from Belgium to France, and when he was visiting his neighboring estates on horseback. Accustomed to his servant-girls, easy conquests as easily forsaken, his mind dwelled upon this unselfconscious young woman who had returned his glance, unsmiling and unabashed. The passing vision of this man on his strawberry roan, with his youthful black beard and romantic pallor, was not unpleasing to the young woman, but by the time he had learned her name she had already forgotten him. He was told that they called her "Sido," short for Sidonie. A stickler for formalities, as are so many "savages," he resorted to lawyers and relations, and her family in Belgium were informed that this scion of gentlemen glass-blowers possessed farms and forest land, and a country house with a garden, and ready money enough. Sido listened, scared and silent, rolling her fair curls around her fingers. But when a young girl is without fortune or profession, and is, moreover, entirely dependent on her brothers, what can she do but hold her tongue, accept what is offered and thank God for it?

So she quitted the warm Belgian house and the vaulted kitchen that smelled of gas, new bread and coffee; she left behind the piano, the violin, the big Salvator Rosa inherited from her father, the tobacco jar and the long, slender clay pipes, the coke braziers, the open books and crumpled newspapers, and, as a young bride, she entered the country house isolated during the hard winters in that forest land.

There she discovered an unexpected drawing room, all white and gold, on the ground floor, but an upper story barely rough-cast and as deserted as a loft. In the stables a pair of good horses and a couple of cows ate their fill of hay and corn; butter was churned and cheeses manufactured in the outbuildings; but the bedrooms were icy and suggested neither love nor sweet sleep.

Family silver engraved with a goat rampant, cut glass, and wine were there in abundance. In the evenings, by candlelight, shadowy old women sat spinning in the kitchen, stripping and winding flax grown on the estate to make heavy cold linen, impossible to wear out, for beds and household use. The shrill cackle of truculent kitchenmaids rose and fell, depending on their master's approach or departure; bearded old witches cast malign glances upon the young bride, and a handsome laundrymaid, discarded by the squire, leaned against the well, filling the air with noisy lamentations whenever the Savage was out hunting.

The Savage—a well-intentioned fellow in the main—began by being kind to his civilized little bride. But Sido, who longed for friends, for innocent and cheerful company, encountered in her own home no one but servants, cautious farmers, and gamekeepers reeking of wine and the blood of hares, who left a smell of wolves behind them. To these the Savage spoke seldom and always with arrogance. Descendant of a once noble family, he had inherited their disdain, their courtesy, their brutality, and their taste for the society of inferiors. His nickname referred exclusively to his unsociable habit of riding alone, of hunting without dog or companion, and to his taciturnity. Sido was a lover of conversation, of persiflage, of variety, of despotic and loving kindness and of all gentleness. She filled the great house with flowers, white-washed the dark kitchen, personally superintended the cooking of Flemish dishes, baked rich plum cakes and longed for the birth of her first child. The Savage would put in a brief appearance between two excursions, smile at her and be gone once more. He would be off to his vineyards, to his swampy forests, loitering long at the wayside inns where, except for one tall candle, all is dark: the rafters, the smoke-blackened walls, the rye bread and the metal tankards filled with wine.

Having come to the end of her epicurean recipes, her furniture polish and her patience, Sido, wasted by loneliness, wept; and the Savage perceived the traces of tears that she denied. He realized confusedly that she was bored, that she was feeling the lack of some kind of comfort and luxury alien to his melancholy. What could it be?

One morning he set off on horseback, trotted forty miles to the county town, swooped down upon its shops and returned the following night carrying, with a fine air of awkward ostentation, two surprising objects destined for the delight and delectation of his young wife: a little mortar of rarest marble, for pounding almonds and sweetmeats, and a cashmere shawl.

I can if I like still make almond paste with sugar and lemon peel in the now cracked and dingy mortar, but I reproach myself for having cut up the cherry-colored shawl to make cushion covers and vanity bags. For my mother, who had been in her youth the unloving and uncomplaining Sido of her first and saturnine husband, cherished both shawl and mortar with sentimental care.

"You see," she would say to me, "the Savage, who had never known how to give, did bring them to me. He took a lot of trouble to bring them to me, tied to the saddle of his mare, Mustapha. He stood before me holding them in his arms, as proud and clumsy as a big dog with a small slipper in his mouth. And I realized there and then that in his eyes his presents were not just a mortar and a shawl. They were 'Presents,' rare and costly things that he had gone a long way to find; it was his first unselfish effort—and his last, poor soul—to amuse and comfort a lonely young exile who was weeping."

Jealousy

"There's nothing for dinner tonight. Tricotet hadn't yet killed this morning. He was going to kill at noon. I'm going myself to the butcher's, just as I am. What a nuisance! Why should one have to eat? And what shall we eat this evening?"

My mother stands, utterly discouraged, by the window. She is wearing her house frock of spotted sateen, her silver brooch with twin angels encircling the portrait of a child, her spectacles on a chain and her lorgnette suspended from a black silk cord that catches on every door key, breaks on every drawer handle, and has been reknotted a score of times. She looks at each of us in turn, hopelessly. She is well aware that not one of us will make a useful suggestion. If appealed to, my father will reply, "Raw tomatoes with plenty of pepper."

"Red cabbage and vinegar," would have been the contribution of my elder brother, Achille, whose medical studies keep him in Paris.

My second brother, Léo, will ask for "A big bowl of chocolate!" and I, bounding into the air because I so often forget that I am past fifteen, will clamor for "Fried potatoes! Fried potatoes! And walnuts with cheese!"

It appears, however, that fried potatoes, chocolate, tomatoes and red cabbage do not constitute "a dinner."

"But why, Mother?"

"Don't ask foolish questions!"

She is absorbed in her problem. She has already seized hold of the black cane basket with a double lid and is about to set forth, just as she is, wearing her wide-brimmed garden hat, scorched by three summers, its little crown banded with a dark-brown ruche, and her gardening apron, in one pocket of which the curved beak of her secateur has poked a hole. Some dry love-in-the-mist seeds in a twist of paper in the bottom of the other pocket make a sound like rain and fingernails scratching silk as she walks.

In my vanity on her behalf, I cry out after her, "Mother, do take off your apron!"

Without stopping, she turns toward me that face, framed by its bands of hair, which looks its full fifty-five years when she is sad and when she is gay would still pass for thirty.

"Why on earth? I'm only going to the Rue de la Roche."

"Can't you leave your mother alone?" grumbles my father into his beard. "By the way, where is she going?"

"To Léonore's, for the dinner."

"Aren't you going with her?"

"No. I don't feel like it today."

There are days when Léonore's shop, with its knives, its hatchet, and its bulging bullocks' lungs, pink as the pulpy flesh of a begonia, iridescent as they sway in the breeze, delight me as much as would a confectioner's. Léonore cuts me a slice of salted bacon and hands me the transparent rasher between the tips of her cold fingers. In the butcher's garden, Marie Tricotet, though born on the same day as myself, still derives amusement from pricking the unemptied bladders of pigs or calves and treading on them to "make a fountain." The horrid sound of skin being torn from newly killed flesh, the roundness of kidneys— brown fruit nestling in immaculate padding of rosy lard—arouse in me a complicated repugnance that attracts me while I do my best to hide it. But the delicate fat that remains in the hollow of the little cloven pig's trotter when the heat of the fire bursts it open, that I eat as a wholesome delicacy! No matter. Today, I have no wish to follow my mother.

My father does not insist, but hoists himself nimbly onto his one leg, grasps his crutch and his stick, and goes upstairs to the library. Before going, he meticulously folds the newspaper, *Le Temps,* hides it under the cushion of his armchair, and thrusts the bright-blue–covered *La Nature* into the pocket of his long overcoat. His small Cossack eye, gleaming beneath an eyebrow of hempen gray, rakes the table for printed provender which will vanish to the library and be seen no more. But we, well trained in this little game, have left him nothing to take.

"You've not seen the *Mercure de France?*"

"No, Father."

"Nor the *Revue bleue?*"

"No, Father."

He glares at his children with the eyes of an inquisitor.

"I should like to know who it is, in this house, who . . ."

He relieves his feelings in gloomy and impersonal conjecture, embellished with venomous expletives. His house has become *this* house, a domain of disorder, wherein *these* "base-born" children profess contempt for the written word, encouraged, moreover, by *that* woman.

"Which reminds me, where is that woman now?"

"Why, she's gone to Léonore's, Father."

"Again?"

"But she's only just started."

He pulls out his watch, consults it as though he were going to bed and, for want of anything better, grabs a two-day-old copy of *L'Office de publicité* before going up to the library. With his right hand he keeps a firm grip on the crosspiece of the crutch that acts as a prop for his right armpit: in his other he has only a stick. As it dies away, I listen to that firm, regular rhythm of two sticks and a single foot, which has soothed me all my childhood. But suddenly, today, a new uneasiness assails me, because for the first time I have noticed the prominent veins and wrinkles on my father's strikingly white hands, and how the fringe of thick hair at the nape of his neck has faded just lately. Can it really be true that he will soon be sixty years old?

It is cool and melancholy on the front steps, where I wait for my mother's return. At last I hear the sound of her neat little footsteps in the Rue de la Roche, and I am surprised at how happy it makes me feel. She comes around the corner and down the hill toward me, preceded by the dog—the horror from Patason's—and she is in a hurry.

"Let me pass, darling! If I don't give this shoulder of mutton to Henriette to roast at once, we shall dine off shoe leather. Where's your father?"

I follow her, vaguely disturbed for the first time that she should be worrying about my father. Since she left him only half an hour ago and he scarcely ever goes out, she knows perfectly well where he is. There would have been more sense, for instance, had she said to me, "Minet-Chéri, you're looking pale. Minet-Chéri, what's the matter?"

Without replying, I watch her throw off her old garden hat with a youthful gesture that reveals her gray hair and her face, fresh-colored, but marked here and there with ineffaceable lines. Is it possible—why, yes, after all, I am the youngest of us four—is it possible that my mother is nearly fifty-four? I never think about it. I should like to forget it.

Here comes the man on whom her thoughts are centered. Here he comes, bristling, his beard tilted aggressively. He has been listening for the bang of the closing front door, he has come down from his eyrie.

"There you are! You've taken your time about it."

She turns on him, quick as a cat.

"Taken my time? Are you trying to be funny? I've simply been there and come straight back."

"Back from where? From Léonore's?"

"Of course not. I had to go to Corneau's for—"

"For his sheep's eyes? And his comments on the weather?"

"Don't be tiresome! Then I had to go and get the black-currant tea at Cholet's."

The small Cossack eye darts a piercing look.

"Ah, ha! At Cholet's!"

My father throws his head back and runs his hand through his thick hair that is almost white.

"Ah, ha! At Cholet's! And did you happen to notice that he's losing his hair and that you can see his pate?"

"No. I didn't notice it."

"You didn't notice it? No, of course not! You were far too busy making eyes at the popinjays having a drink at the café opposite, and at Mabilat's two sons!"

"Oh! This is too much! I, I making eyes at Mabilat's sons! Upon my word, I don't know how you dare! I swear to you I didn't even turn my head in the direction of his place. And the proof of that is . . ."

Indignantly my mother folds her hands, pretty still though aging and weatherbeaten, over a bosom held up by gusseted stays. Blushing beneath the bands of her graying hair, her chin trembling with resentment, this little elderly lady is charming when she defends herself without so much as a smile against the accusations of a jealous sexagenarian. Nor does he smile, either, as he goes on to accuse her now of "gallivanting." But I can still smile at their quarrels because I am only fifteen, and have not yet divined the ferocity of love beneath his veteran eyebrow, and the blushes of adolescence upon her fading cheeks.

So little is known of the Russian poet Marina Tsvetaeva (or Cvetaeva) in English, and she is so apparently a major writer, that I have included, along with a brief poem (very little of her work has yet been translated into English), a moving letter which has been kindly provided, translated, and annotated by Professor Simon Karlinsky, whose book Marina Cvetaeva: Her Life and Art (Berkeley, 1966) introduced me to this remarkable and tragic woman, one of the handful of great women writers to have had a family—she had three children, one of whom died of starvation in Moscow during the civil war. In 1921 she left Russia with her surviving daughter, to whom she was greatly attached, and her husband, Sergei Efron, an officer in the White army, and lived in great poverty in Berlin, Prague, and Paris, giving birth to a son in exile. In 1939 she returned to the Soviet Union, again following her husband, who had become a double agent and who believed that he would receive clemency from Stalin; her son also longed to become a part of the country he had never seen. However, Efron was executed at once, and when the family returned, the daughter was sent to a labor camp and Tsvetaeva was alone in Moscow, unable to earn a living. When the war began, she was evacuated with her son to the little town of Elabuga, where she could get work only as a kitchenmaid. It was there, "in a state of acute mental depression and utter hopelessness," * that she hanged herself on August 31, 1941.

As emphasis to Tsvetaeva's letter, here is one by Katherine Mansfield, quoted by Tillie Olsen in her essay "Silences: When Writers Don't Write":

> The house seems to take up so much time . . . I mean when I have to clean up twice over or wash up extra unnecessary things. I get frightfully impatient and want to be working. So often this week you and Gordon have been talking while I washed dishes. Well someone's got to wash dishes and get food. Otherwise "there's nothing in the house but eggs to eat." And after you have gone I walk about with a mind full of ghosts of saucepans and primus stoves and "will there be enough to go around?" And you calling, whatever I am doing, writing, "Tig, isn't there going to be tea? It's five o'clock."

* Karlinsky, p. 105.

MARINA TSVETAEVA
(1892–1941)

COMMENTARY AND TRANSLATION
BY SIMON KARLINSKY

This letter was written from France by the great Russian poet Marina Tsvetaeva to her Czech friend the journalist Anna Tesková, whom she came to know during an earlier stay in Prague. In the 1920s Tsvetaeva was at the peak of her creative powers as a poet, playwright and literary critic. But she, her husband, her grown-up daughter and her little son lived in extreme poverty, and the poet constantly found herself drowned in housework and in the endless chores of supporting a family of four. Her husband had no permanent job and spent much of his time dabbling in esoteric émigré politics (Eurasianism). The two-year-old boy demanded constant attention. All the cooking, washing, laundering and marketing were done by Marina. In the midst of all this, she managed to write several profound metaphysical verse plays and some of the most beautiful and original poetry ever written in Russian. Her letters to Anna Tesková, published by the Czechoslovak Academy of Sciences in 1969, tell us for the first time how she felt about it all.—S.K.

Meudon, Dec. 12, 1927

. . . It will soon be Christmas. To tell you the truth, I've been driven so hard by life that I feel nothing. Through these years [1917–1927] it was not my mind that grew numb, but my soul. An astonishing observation: it is precisely for feeling that one needs time, and not for thought. Thought is a flash of lightning, feeling is a ray from the most distant of stars. Feeling requires leisure; it cannot survive under fear. A basic example: rolling one and a half kilos of small fishes in flour, I am able to think, but as for feeling—no. The smell is in the way. The smell is in the way, my sticky hands are in the way, the squirting oil in the way, *the fish* are in the way, each one individually and the entire one and a half kilos as a whole. Feeling is apparently more demanding than thought. It requires all or nothing. There is nothing I can give to my own [feeling]: no time, no quiet, no solitude. I am always in the presence of others, from seven in the morning till ten at night, and by ten at night I am so exhausted—what feeling can there be? Feeling requires *strength*. No, I simply sit down to mend and darn things: Mur's, S.'s, Alya's,* my own. Eleven o'clock. Twelve o'clock. One o'clock. S. arrives by the last [subway] train, a brief chat, and off to bed, which means lying in bed with a book until two or two-thirty. The books are good, but I could have written even better ones, if only . . .

It is a fault (it is nobody's fault), perhaps of my own. Except for nature (i.e., the soul) and the soul (i.e., nature), nothing touches me—not the social issues, not technical [progress], neither —— nor ——. Therefore, I hardly go anywhere—it is such a bo-o-o-re! A professor is lecturing and I'm counting: how many minutes till the end? Why bother? Just like today: a Eurasianist lecture on linguistics. It would seem the topic is one close to me. It only seems. The professor (he is a celebrity) derives all languages from four words. When I heard that, I was repelled at once: nothing good can come from anything that *even-numbered*. What about rhyme? Rhyme is a *third* [dimension]! So I did not go and am sitting here between a stocking and the ticking alarm clock.

How I long for Prague! Will it ever come true? Even if *not*, please say *yes!* Never in my life did I yearn to come back to a city; I have no wish at all to return to Moscow (sooner anywhere else in Russia!), but I yearn for Prague; it must have transfixed me and cast a spell over me. I want *the myself* of those days, the unhappy-happy myself of *The Poem of the Hill* and *The Poem of the End*,† myself as a disembodied soul of all those bridges and places. Here are those verses:

> Myself—a disembodied soul
> Of all those bridges and places.
>
> Where I lived and sang
> Alone, like a spirit, like a [flag]pole,

---

* Mur was Tsvetaeva's son; S. was her husband, Sergei Efron, shot in 1939 after he returned to the U.S.S.R.; Alya is their daughter, Ariadna Efron, aged fifteen in 1927, who now frequently edits publications of Tsvetaeva in the Soviet Union.—S.K.

† *Poem of the Hill* and *Poem of the End* are two long narrative poems Tsvetaeva wrote in Prague in 1924–25. Both deal with a heartbreaking affair she had there with a former White Russian officer. *Poem of the End* is a masterpiece of in-depth psychology and an astounding feat of verbal inventiveness. It is now regularly reissued in the Soviet Union and has tremendous prestige among the younger generation of Soviet writers.—S.K.

> I want myself, a disembodied soul
> Of all those bridges and places.

That is how I could write verse once; I've written no individual, separate poem since 1925, in the month of May. "You may not!" is my private verb.

How I'd love to go with you—all over Czechoslovakia. Into the depth of it! (I know it is utterly unrealizable.) After Prague (a city of visions), into nature. Can it all really be the matter of money, which, whether I had it or not, I've always despised? Yes, had I the money . . . Teachers and books for Alya, a reliable nanny for Mur, an apartment with a little garden—and for myself? To be able to write every day and to travel twice a year, and the first trip would be to see you. Someone told me recently that the hundred thousand rubles I once had in Russia would now amount to a million francs. A [mere] sound.

I have finished *Phaedra*. It took six months to write, but, then, I can write for a half hour, at the very most an hour a day. It is quite long, longer than "Theseus." I plan to expand *Theseus* into a trilogy: "Ariadne," "Phaedra" and the still-to-be-written *Helen*.* I do not know where I'll submit it. Perhaps *The Contemporary Annals*. Now I'm tidying it up and editing. There are many places where I did not quite bring it off. I know I'll manage. It is my notebook that keeps me above the surface of the water.

Goodbye! Think of me on the bridges and the narrow streets of Prague. Do you suppose we'll ever get together? . . . [Next ten lines deleted by the editor of Tsvetaeva's letters.]

---

* *Theseus* was a trilogy of verse plays, in which Tsvetaeva modernized the Greek myth and traced three kinds of involvement between women and men. *Ariadne* shows a poetically-heroic, intellectual encounter between Ariadne and the young Theseus. *Phaedra* is about the desperate and irrational passion of an older woman for a hostile young man. Because of the adverse reaction of Russian émigré critics to *Phaedra* after it was printed in the prestigious journal *The Contemporary Annals*, Tsvetaeva never completed the projected third part about the tragic involvement of the aged Theseus with the girl-child Helen (the future Helen of Troy). The first two plays, ignored when first published, are now considered masterful and innovative works, unlike anything else in Russian drama.—S.K.

In Praise of Aphrodite

The old gods are no longer bountiful
On shores that no more edge that ancient stream.
Fly, doves of Aphrodite, far away
Into the wide gates of the setting sun.

Lying on sands that are already cold,
I pass beyond into a dateless day. . . .
Like a snake staring at its shriveled skin,
I know that I have left my youth behind.
(*translated by George L. Kline*)   1921

————————

"Kindness" is one of the poems Sylvia Plath wrote in the last few months of her life, rising at 4 A.M. in her London flat to work before her two small children awoke, writing when she shared the dark world only with the milkmen.

SYLVIA PLATH
(1932–1963)

Kindness

Kindness glides about my house.
Dame Kindness, she is so nice!
The blue and red jewels of her rings smoke
In the windows, the mirrors
Are filling with smiles.

What is so real as the cry of a child?
A rabbit's cry may be wilder
But it has no soul.
Sugar can cure everything, so Kindness says.
Sugar is a necessary fluid,

Its crystals a little poultice.
O kindness, kindness
Sweetly picking up pieces!
My Japanese silks, desperate butterflies,
May be pinned any minute, anesthetized.

And here you come, with a cup of tea
Wreathed in steam.
The blood jet is poetry,
There is no stopping it.
You hand me two children, two roses.

Scenes from Family Life

The incredible complexity of family relationships! The basis of many novels, from Tolstoy's War and Peace to Mann's Buddenbrooks and Compton-Burnett's Delphic fiction, it is especially difficult to capture in short fiction (and to chop off an excerpt from a long, dense work would leave the blood spurting out in all directions), but Donald Windham has done it superbly in The Starless Air, recreating with discreet artistry those interminable family holidays to which we are bound by our earliest experiences, like them or not, and to which we keep returning—in memory, if not in the flesh. The web of the family: how it comforts and shields, how it would keep us children forever! The center of most women's lives, it is a territory as perilous as any battlefield, as mysterious as any deep cave. This particular sense of hiddenness and mystery, of each family member leading his own interior life within the close confines of the home, is developed beautifully by Katherine Mansfield in her Prelude and At the Bay, which would have been included but for lack of space.

DONALD WINDHAM
(1920–   )

The Starless Air

Quivi sospiri, pianti, e alti guai
risonavan per l'aer senza stelle.

The house was painted dark green; but, like the dark green leaves
of the sterile magnolia tree in the front yard, it was almost black with
the soot of the city. Dark with cynicism and impunity it stood, the only
private residence remaining in a block of boardinghouses and stores
guarded at each corner by a filling station. The ornate and superior
structure of the building had always dominated the neighborhood of
comfortable homes which grew up and declined about it, and now it
added the obscure distinction of being obsolete. A deep veranda ran
the length of the front and halfway back on either side. Over the
front door the original street number, 910, appeared in a semicircle of
colored glass. Peachtree Street had been renumbered years ago and
the new address was nailed in metal numerals on one of the square
columns which, with their connecting balustrades, enclosed the porch;
but the family still referred to the homeplace by its original number.
Above and beyond the slate roof of the porch, half a dozen second-
story windows, each surmounted by a semicircle of colored glass,
flanked a small upstairs veranda obscured by a wooden spiderweb of
knobbed spindles. The third story was used only as an attic, but it also
opened onto two balconies: one was bannistered on three sides by
diamond-shaped openings, the other forced upward through a series
of oval frameworks into a small turret which reached above the top
branches of surrounding oak trees.

The hall was cold. The house had been built for utility and comfort; it had served as the center of life for a large family where meals
were eaten and comfortable evenings spent, but it had no furnace. No
furnace had been needed in the days when fires were built in all the

rooms and their heat warmed the corridors. But now that only a few rooms were used, the hall was icy.

A woman whose face was puckered but relaxed with a baffled resignation came out of the living room and closed the door. She hurried to the back end of the cold hall and passed, again closing the door behind her, into a one-story wing which ran behind the main square of the house alongside of a grassless stretch of ground where Papa's hunting dogs had been kept. The fence was gone from around the dog yard and only one tarpaper-roofed doghouse was visible in the gray morning twilight which greeted her through the dining-room window as she entered. The fire burned brightly in the room, however, and she knew that her sister was downstairs. She pushed through the first swinging door into the pantry and through the second into the kitchen.

"Merry Christmas, Hannah," she cried. "I didn't know you were down already."

"Merry Christmas, Lois," Hannah answered, turning from the fire which she was lighting in the wood stove. "I thought we'd get an early start this morning, but there's no sign of that Mamie."

Hannah was tall and thin with a thornlike face and figure and a sharp voice. She was the oldest of four sisters and contrasted to Lois who was the youngest and whose plump, fruitlike figure was not yet accustomed to the weight it was adding. With two children each, they lived alone in the large house. They ate together and shared the dining room and kitchen; otherwise, they lived as separately as possible. Lois *had* the downstairs bedrooms and living room. Hannah *had* the parlor and library and two bedrooms upstairs. The rest of the house was empty.

"I don't think I'll light the gas oven till we start cooking," Hannah said. "It's warm enough this morning. Mamie doesn't even have the excuse of bad weather."

"No, we needn't have wrapped the pipes in crocus sacking if the weather keeps up like this," Lois agreed.

"Well, it doesn't hurt to be on the safe side," the older sister said.

"No," the younger replied. "And it'll be nice to be able to have the children play outside."

Hannah, who had gone to the sink to wash her hands, frowned.

"Mine are still asleep."

"Well, they've outgrown Santa Claus," Lois said. Then, suddenly remembering their brother, she asked: "What about Bobby?"

Hannah clicked her tongue against the roof of her mouth.

"He's still asleep. I looked in and covered him up. He'd kicked all the covers off onto the floor."

"I declare, it's a wonder they don't catch pneumonia when they're like that," Lois lamented. "I'm glad Mama didn't live to see him do it."

"Well, if Mama were still alive," Hannah said, going to the little pantry and coming back with her arms full of meat, "you can be sure that he wouldn't dare turn up here that way. Christmas hasn't been the same since Mama died."

"No," Lois agreed. "There's less spirit. It seems like there's less of everything but people, and every year there's more of them. How many are coming today?"

"I haven't stopped to count," Hannah said as she poked a fat turkey with her bony finger.

"Let's see," Lois suggested and held up her hands to count on. "There's nine of us brothers and sisters, two aunts and Cousin Ella, my two children, your two, Bobby's five, Ivy's husband and two children, Audrey's husband and Fred junior, Henry's wife and child and Stuart's wife and child. Why, that's thirty in all."

"You've forgotten Cy's wife," Hannah said.

"Yes, Cy's wife. Thirty-one. You know, I don't believe there were ever more than two dozen when Mama was alive. But you know she'd be happy to know we've all stayed together.

"Do you know where the onion board is?" Hannah interrupted.

She could not start the turkey dressing without the onion board and they were still looking for it when the doorbell rang.

"There's the doorbell," Hannah said. "Can you go? It might be Audrey, though I don't know how she ever got away from that Fred Bronson this early."

It was Audrey. Lois let her in, and both of them had to make two trips to the car to carry in all the cakes which she had brought. Her husband, who had driven her over, would not get out of the car.

"Who's staying with Fred junior?" Lois asked.

"That Eustasia," Audrey answered, pronouncing the name as though it were a foreign word. "I don't understand her or any of Fred's sisters. She's just as sweet as sugar, now, though."

"Isn't that the way," Lois said.

She set the huge fourteen-pound fruitcake down on the table beneath the full-length life-sized portrait of Papa so she could close the front door. When she entered the kitchen with the cake, Hannah, who had found the onion board where it usually stayed, pointed at the roasting pan in which the cake was cooked and announced that that was what she wanted.

"Now I can finish this dressing and get the turkey in the oven."

"Oh, I meant to tell you," said Lois, "the all-blue lights look real nice on your tree."

"I think they're nice," Hannah agreed. "That's the way the Stantons do their tree now."

"Oh, is that so?"

The Stantons were the family of Hannah's husband from whom she took all her uppity ways, and Lois did not see why Hannah spent so much time with them when her husband had been dead ten years.

Hannah was mixing the sharp oyster dressing for the turkey, Audrey was soaking sweet wine in the fruitcake and loosening it to turn out of the pan, Lois was cutting up fragrant pineapple, oranges and dates for the ambrosia, when Mamie arrived. The Negro stood in the open doorway letting the fresh air blow into the kitchen and stir the rich odors.

"Christmas gift, Miss Hannah! Christmas gift, Miss Lois! Christmas gift, Miss Audrey!"

Audrey shivered and cried, "I'll Christmas gift you, Mamie, if you don't come in here and shut that door. I'm freezing already."

Hannah and Lois exchanged a look over their sister's head. Audrey's steam-heated house was so hot that they almost suffocated each time they went to see her.

"It'll be warm as toast in here in a little," Lois said. "Shut the door, Mamie, and come on in. There's plenty for you to do. The children haven't had their breakfast yet."

Mamie went into the servant's room back of the kitchen to change her clothes. After the gust of fresh air, the kitchen was more redolent of food than before. Lois put the ambrosia in the icebox in the little pantry and came back into the kitchen for the bowl in which to mix the dough for the rolls. By habit and without looking, she reached her hand to the shelf by the red flaking wall and found it. There was enough cooking on a routine day to keep the sisters in the kitchen all the time they were not at housekeeping, cleaning rooms, making beds, sorting dirty clothes in the large wicker basket for the wash-woman. The familiar dun-colored plaster of the kitchen walls showed through where the paint had cracked and peeled away and it seemed to be imbued with moisture like the steps at the end of the porch where the children had tried to start a fire one day by holding lit matches to the wood, which had been so wet and earthlike it had only smoked. Within them, the routine of eating kept the air forever dark and whirling, like a whirlpool filled with dust, and no violence which happened outside the house could prevent the sisters from concentrating blindly on their own existence.

"Just let me show you something," Lois said, standing in the middle of the room.

In one hand she held the bowl. With the other she lifted her skirt to reveal a purple-and-green bruise on the white fat-scarred flesh of her thigh. Audrey, whose body, widowed in a different way, was also beginning to expand in a vegetable growth without resemblance to the human skeleton, gasped in commiseration.

"Why, Lois, however in the world?"

"Bobby."

"Was he . . . ?"

She formed the last word with her lips but did not speak it.

"Yes."

"Did he pinch you?"

"No," Lois said, dropping her dress. "He came in last night and left a box in the dark hallway and I fell over it. I was so mad at him I could have killed him."

"He might have broken her leg," Hannah added from the stove.

"Isn't it the truth," Audrey agreed.

"But I had to forgive him when I saw what it was. A box of toys for his children. When I saw that even in that condition he had remembered them, I had to forgive him."

"Bobby's not bad," Audrey said. "He never acted like that when Margaret was alive."

"Well, what do you reckon makes him do it now?" Lois asked, shaking her head. "Those poor children."

She had gone into the big pantry from which she called back these last words. The big pantry, lined with shelves on which utensils and staple foods were stored, smelled like a musty gourd; and somewhere on one of the top shelves was stored the gourd dipper which had been used at the well when they were children. She pushed the biscuit board from the top of the flour barrel and reached down to fill the sifter with flour.

"Let me tell you something I promised not to tell," Audrey said, standing halfway between Hannah at the stove and Lois in the pantry. She put her hand on the back of a chair with a hairy cowhide seat and waited until her sisters stopped to listen.

Lois knew that Audrey had always had a temper and she heard the familiar tone of anger in Audrey's voice. She remembered the night when Audrey had run away in a fit of temper and married because Mama had scolded her at a party. They said that her husband's bad blood had been responsible for her invalid child. Lois forgave Audrey her temper whenever she remembered that Audrey was trapped in life with that Fred Bronson and had to spend all of her days tied to him.

"I went down to Bobby's house last week to try and get him straightened out," Audrey said dramatically. "And the house—the house was such a mess that I don't see how those children live in it."

"It shouldn't be," Lois objected. "They've got Edith there all the time with nothing to do but take care of them."

"She's just a good-for-nothing *nigger*," Hannah whispered, looking toward the back room.

"Anyway," Audrey went on, "ever since Margaret died I've said that what Bobby needs is a job to keep him occupied and I decided to do something about it. I went down there and sent him over to see Cy."

At the name of their oldest brother the other two sisters both tried to speak at once.

"A fine older brother *he* is," Hannah cried, drowning out Lois. "He hasn't been to see us for months. And when I called to see if he was coming to Christmas dinner and to ask him to bring the kick for the eggnog, he said that he would but that he'd have to leave early because he is going to have a drink with the mayor. The *mayor*, mind you."

"Wait till you hear the rest of this," Audrey said. "After all, Bobby is Cy's brother, his own flesh and blood. He ought to help him even if nobody else will. And he's in charge of the hiring for all the construction work the city does."

"It's a shame none of the boys but Cy finished college," Lois sighed.

"Well, that wouldn't have helped much in this case. Do you know what your brother Cy had the nerve to tell Bobby?"

Audrey paused and her sisters looked at her gravely.

"He said that there is a city ordinance which forbids anyone hiring any of his own relatives."

"It's a city ordinance I never heard of," Lois admitted.

"It's a city ordinance *no* one ever heard of," Audrey hissed.

"I'll bet he's got plenty of *her* relatives on the payroll," Hannah whispered and leaned her head forward. "I'll bet there's no law against *that*."

"Well, we'll just have to call Cy aside while he's here today and tell him that he has to do something about Bobby," Lois said. "If he can't give him a job let him advance him enough money from the body of the estate to set him up in some kind of business. A filling station, or something."

"But do you think he'll do it?"

"If we tell him to do it, he'll have to," Audrey stated. "You speak to him. I'm not speaking to him, but you speak to him and I'll stand behind you."

The sisters agreed on this, and Lois went into the back room to see what in the world was keeping Mamie. Mamie's mother had lived in the back room; but, for some reason the sisters could not understand, Mamie preferred to live in a shack in Niggertown; and the room was now used for storage. Its odor of old newspapers stacked on trunks and broken leather and horsehair couches was pierced by the sharp scent seeping from an earthenware jar of home brew in the corner. Sitting on a bottomless chair across which a stack of newspapers was folded, Mamie was looking through her pocketbook. Lois stared at her and demanded:

"Mamie, what are you doing? Come on in here and cook the children's breakfast before you freeze and they starve."

She returned to the pantry and hurriedly made the rolls which would have to set and rise. In the kitchen she heard Audrey ask

Mamie what she thought of a man who would not give his own brother a job, and heard Mamie laugh and answer:

"Yessam."

When the children had eaten, Lois and Mamie cleared the dining-room table to put in the extra leaves. With them, the round table extended into an oval which filled the room from fireplace to sideboard and from door to windows. Before they finished, the doorbell rang and Bobby's children arrived. Lois let them in. Neat and uncomfortable with respect for 910, they stood in the hall and assured her that they had eaten breakfast, so Lois told them that their father was still asleep and led them into the living room to get their presents. She was glad that their presents were waiting for them under her tree rather than under Hannah's. Hannah had the fine rooms and received the grownups, she told herself, but it was Aunt Lois that the children liked and to her side of the house that they came.

The living room was warm with the odor of pinetree. She had lit the fire and watched her children, in their bathrobes and bedroom slippers, open their presents before she had gone back to the kitchen to help Hannah. They were dressed now. Her oldest boy was standing by the tree before one of the six windows which looked out onto the front porch; but the youngest was lying on the floor by the fireplace arranging colored marbles along the pattern of the hot rug. She sat down and asked him to come and sit in her lap for a minute. As he climbed up she felt his forehead and thought that he had a fever. A terrible fear of death and the outside world swept over her. Growing up in the comfortable life of the homeplace had not bred in her the faculty of resisting. She accepted the inevitable. Together with her brothers and sisters, she felt no capacity or desire to extend the limitations of her existence; she desired only to keep what she already had. She would rather cut off her right arm, she often said, than have to choose between her two boys or have anything happen to either of them. But as she sat by the glowing fire stroking her son's forehead she realized that he was merely hot from the fire and if he was pale it was probably that his stomach was upset from overeating. Her fears subsided, but she promised herself that she would look at him again in a little while and if he was still pale she would put him to bed.

When she reached the dining room Mamie had set the table with the everyday silver.

"Mamie! What do you think I made you polish the good silver for yesterday?"

She sent Mamie back to the kitchen to wash the cake stands and she set the table herself. She was thinking about Bobby's children as she turned from the table to go to the kitchen and she was startled to see Bobby leaning in the hall doorway watching her.

"For goodness' sake, Bobby," she gasped, "you startled me."

Her brother leaned in the doorway without replying and smiled at her. His large-pored red face was bloated beneath his stubble of brown-silver beard, his small blue eyes were bloodshot, and even at this distance she could see that his smile was intoxicated. Slowly he heaved his weight away from the doorframe and entered the room like a falling object. He came to rest against the sideboard and Mama's cutglass cups rang perilously against the side of Mama's cutglass punchbowl. His smile widened with defiance and he launched himself off again from the sideboard toward Lois. She could have dodged, but she knew that if she did he would land on the table and break Lord knows what, so she stood still and let him land with his arms about her waist. The reek of sweat and whisky breath engulfed her.

"Bobby, wherever did you get a drink in this house at this hour of the morning?" she asked in despair.

For answer he pinched her thigh.

"Bobby! Your children are here. You don't want them to see you like this. Come on back upstairs with me. Audrey's in the kitchen. You don't want her to see you like this."

"Where's Audrey?" he shouted and let go of Lois. "She loves her brother."

For a second he stood on his own. Then his equilibrium collapsed and he careened through the swinging door into the little pantry. He bumped into the right side of the first doorframe and into the left side of the second. There the door swung open and he stood facing two frozen gazes of admonition. He replied with a half completed gesture, aware of the impossibility of his situation but unable to escape it. His face puckered between crying, shouting, and remaining silent; then he produced a bottle of whisky out of his trousers pocket and whispered:

"Once a year, remember? Have a friendly little drink with your brother Bobby."

With an insufficient disguise of dryness, Audrey said:

"All right, Bobby, give me the bottle and I'll pour us a drink."

The other sisters caught on. Hannah stood firm while Audrey advanced and Lois furnished a rear guard. Their brother stumbled forward in momentary excitement and laughter and flung his arms in a bear hug about Audrey's head as he bit her neck. She screamed and struggled to escape. Undeceived, he held her fast and with his hands free behind her turned up the bottle to his mouth.

"Ha, ha, ha," he laughed as the whisky gurgled and overflowed his mouth, ran through the beard along his chin, and dribbled onto his sister's back.

He twisted and turned, but the sisters had him surrounded and as he turned from Hannah he faced Lois who wrenched the bottle from his grasp. He reached after her, but Audrey was still in his arms and hindered his pursuit. Lois disappeared through the dining room into the hall. She was almost to the front when she remembered that the children were there. She turned and retraced her steps to the back where she entered the bedroom.

The shade was down and the room dark. Faint glows illuminated the pinpricks in the green cloth through which the sun shot in shafts on bright days. Panting and listening to her heart beat, she leaned against the heavy carved foot of the bed; but her breath would not come slowly nor her heart cease pounding. She reached out one hand and raised the shade. The day was so dark that no more light entered and she lowered the shade again. The reek of man—whisky, smoke, and sweat—had brought back all the melodrama of marriage in the barren landscape of Florida. Her nostrils cringed, and looking down she saw the open bottle clutched in her hand. She crossed to her wardrobe trunk which stood open with its back sides against the wall. Its drawers and closet, covered in a pattern of bluebirds flitting in the branches of trees, faced out into the room; and out of sight, down behind the shoe rack on the closet side, she set the bottle. Mama had given her the trunk as a wedding present. That had been in a different world, peopled with different people. In it had been only pleasure; from it had been banned all which demanded decisions. She remembered sitting on the front porch with Cousin Ella the first time she saw her husband. It was summer. She did not remember the month, but he was dressed all in white and in his hand he carried a small black case like a portable typewriter. (A Vanaphone—they ceased to exist long ago.) He came smiling up the walk toward her, as handsome as could be, and he asked her mother if he could call on her the next day. Till late that night she read *Elsie Dinsmore*, the only book she loved. She remembered leaving through the white picket fence and down the tree-lined dirt street on her honeymoon. Melodrama and guilt faded with the Florida sky. Reality and innocence returned with her return to the homeplace. Here were no interminable oceans, no brick walls, only the lightning vine spreading across the vacant lot which had been the pasture. The picket fence was gone, the street paved. An electric trolley roared past beneath the overarching trees. She borrowed money from Mama and paid the bad debts her husband had left in all the towns through which they had passed in five years. Mama said that she need not ever leave the homeplace again. But the homeplace was a different world. Papa died and Mama died. At each funeral she sent the children down to the back end of the vacant lot where the cherry trees had bloomed when she was a girl. She did not want them to feel sorrow. The people she had known moved from the neighborhood. Autumn after autumn, her brothers married and moved from the house, leaving their rooms empty. Spring after spring, they returned, alone, in the middle of the night, drunk, seeking the immunity they had once known, falling asleep across her bed and vomiting in her shoes.

She turned away from the dark cloth of bluebirds and stood up. Before she left the room she opened the small top drawer of the trunk where she kept her scarfs and summer handkerchiefs. In a little lidded compartment at the end, among her rings, earrings, pearshaped rhinestone buttons and glass beads, she found a small black bankbook and looked in it. For many months of all the years since she had returned home, the figures on the pages, varying in the first numeral, repeated insistently the three noughts. This reassured her, and she went to the living room to see how her son was.

The other children were in the front yard playing; she could see them through the windows; but he still lay before the fire. Frightened by what she had felt before, she made him lie down on the couch and tucked a blanket about him. He faced toward the fire and could not see the Christmas tree behind him, so she hung the two stockings bulging with apples, oranges, toy automobiles and hard candies, back over the mantel, one anchored beneath each front leg of the big clock.

"Who is that?" he asked, pointing to an oval gold-framed photograph above him, halfway up the wall toward the high ceiling.

"That's Uncle Todd," she told him as she had many times before.

"He was a twin, wasn't he?"

"Yes. His brother's name was Dunk and they had only one picture taken because they looked so much alike."

"Dunk and Todd are funny names."

"I guess so, Sugar," she said and kissed him.

She was nearly to the dining-room when she heard his voice calling: "Moooother!"

She hurried back up the hall wondering what could be wrong. She stood in the open door asking him while he looked at her without answering. Finally he said:

"I love you."

The meal was progressing and the air was turgid with the aroma of cooking turkey and ham. Mamie was washing vegetables at the sink and Hannah was cutting up potatoes on a marble slab.

"Did you get Bobby back to bed?" Lois asked.

"Audrey's upstairs with him now."

"I don't know what makes him do it. And I think Sonny's sick."

Hannah clicked her tongue on the roof of her mouth in sympathy, and suspiciously watched Lois who had opened the oven door and was basting the ham. Just then the doorbell rang and the sisters sent Mamie to answer it.

"It's Miss Ivy," she said on her return.

The sisters hurried up the hall. The front door stood open, letting in the cold air, and at the steps Ivy was being helped up by her daughter.

"Merry Christmas," she shouted in a voice weak with laughter.

"All right, Mama, you've got to step up just a little," her daughter said with amusement. "I'm helping, but you've got to step up just a little."

"Tony," Hannah called to Ivy's husband who was parking the car in the driveway. "Drive on to the backyard so the others can get in."

Tony was Ivy's second husband. They disagreed on everything. When they reached the parlor Ivy sat on the loveseat and he retired to the corner and did not speak except to answer questions. Ivy took charge of the conversation and in a nasal voice repeated what the doctors had told her about her fallen colon. She illustrated with imaginary diagrams on her black and white print dress. When she finished, Audrey came in and she had to repeat it all. Together, the four sisters resembled each other very closely. Their figures varied from Hannah's tall and thin to Ivy's short and fat; but their faces were alike and they all wore their hair the same, in lifeless regular waves with a few carefully disordered curls just behind their ears.

"Does it hurt?" Lois asked.

"Pshaw, no, except I feel so faintified," Ivy laughed. "I don't get to eat hardly a thing."

"Isn't that a shame."

Lois and Audrey took the coats to Lois's bedroom and threw them across the bed.

"Did you get Bobby back upstairs?" Lois asked.

"Yes, but he's not asleep. The Lord only knows when he'll come barging down again."

"I declare. Whatever in the world do you reckon makes him do it?"

When they returned to the parlor, Hannah was telling about Bobby and they joined her in explaining their plan to Ivy. Ivy did not like being told about the plan after it was formulated and her voice was petulant.

"I guess you're right," she said, "but Bobby ought to be ashamed of himself. He doesn't have a worse time than the rest of us. I've been sick for weeks and I'm going to have all these doctor bills."

"But, Ivy, we have to do something about Bobby."

"I admit that," Ivy admitted. "But the way I feel is that the Lord helps those who help themselves."

She struggled delicately to lift her short legs from the floor and sat with them stuck defiantly along the loveseat. Suddenly she turned to her husband with a little shriek.

"Where's Cousin Ella?"

"I left her in the kitchen," he replied.

"Well, for Heaven's sake keep her away from me," she told her sisters. "She's nearly driven me out of my mind."

"Oh, Mama, you shouldn't let her bother you," her daughter said.

"You wouldn't say that if you were helpless in bed and couldn't get away from her," Ivy cried. "I know now why she comes and helps when there's anyone sick or dead. It's not because anyone wants her but because the people she's staying with at the time make sure she goes. Do you know what she was doing this morning? Christmas morning? When the children got up early she was already out, had fixed herself a pot of coffee and drank it, and was sweeping the front yard. And she wouldn't come in, either. I called to her that there was no need for her to do that, but she just went right ahead and finished. And *then* she went around the house and swept the backyard. I get so nervous that I know she's keeping me from getting well, and I don't know what the neighbors think."

"I hope she doesn't find out that Sonny's sick," Lois said. "He's scared of her. Once when she was here I heard her telling him that the reason she never got married was that the only thing she loves is coffee."

"Shush," Hannah shushed them. "I think I hear her coming."

Cousin Ella was thin and wild and wore a long dress made out of a stiff material no longer seen anywhere else except in an occasional square of a crazy quilt. She kissed the sisters and started chattering; but as soon as Hannah said that she had to go back to the kitchen to remind Mamie to put the boiling bacon in the vegetables, as she always forgot, Cousin Ella said that she would help and followed her out.

"Well, thank goodness," Ivy sighed. "But I doubt if we'll ever get any dinner now."

Everyone laughed.

Aunt Lil and Sally arrived next. Lois let them in and led them to the parlor. Everyone shouted at once in greeting, but all voices were lost beneath Aunt Sally's insistent announcement that she had never seen a Christmas day with such weather and that the air outside was a regular brown fog. She moved forward as unobstructably as an elemental force till she reached the comfortable chair near the fire and made Ivy's daughter move so she could sit there.

"So this is Ivy's girl," she said patting her arm. "Honey, you've grown so since last time I saw you I wouldn't have known you. Now run back to the kitchen and do Aunt Sally a favor. Just tell them that you want a coffee can for her and Aunt Lil. They'll know what you mean."

Henry and his wife arrived, then Stuart and his wife. The children stayed in the yard. Lois and Hannah were very polite to the wives for they did not like them. They came from other towns and were not their kind. But, as Hannah said, if their brothers were going to act the way they acted, you couldn't expect any other kind of women to marry them. Jefferson had a very nice bride, but then Jefferson was a very nice boy.

There was a great deal of loud talk in the parlor and Lois told Aunt Sally that Sonny was sick at his stomach. The old woman did not wait for her to finish speaking but shouted with the unquestionable authority of a woman who had outlived a husband, two children and three grandchildren:

"Give him calomel! Tomorrow morning give him a good dose of salts. That's what I always gave my Carl and there wasn't a stronger boy in these parts."

"But calomel makes him nauseated, Aunt Sal . . ."

"Give him calomel, and a piece of Juicy Fruit gum after it and he'll like it. I raised eight younguns and never found a one that didn't like it. Here, let me fix it and give it to him for you."

She placed a hand on each arm of her chair and pushed her body up, grunting loudly. When she was standing she followed the grunts with a long, low whistle.

"It's becoming a powerful effort for me to get around," she complained.

Aunt Lil, who was thin, had sat across the fire in a straight chair.

"The trouble with you, Sally, is you're getting fat," she cried in a dry voice.

She sat very straight and leaned forward from the hips and she shifted the snuff to the other side of her mouth and spat toward the coffee can on the hearth. Two drops of the brown liquid splattered on the ash-dusted hearth like raindrops on dusty earth.

"That's the funny thing," she continued. "I can't understand it. Sally and I both had all our teeth out, and ever since she's been getting fatter and I've been getting thinner. Had just the opposite effects on us, and done to us both just what we didn't need."

"Aunt Lil," Henry said, "what you need is a bottle of Hannah's good home brew."

A few minutes later, Lois and Sally returned to find everyone with a bottle of home brew in his hand. Sally wanted one too.

"Don't give Sally one," Lil screamed. "She'll break that chair when she sits down if she gets any fatter."

"I don't care," Sally replied. "I want one, and I don't care."

"Sally, Sally," Lil cried, standing up, turning round and laughing. "Look how fat it's making me already."

"Remember the last time Anna came to see us," Audrey said through tears of mirth. "Uncle Tom drove her in from the country one Sunday to have dinner with us, and when they got here she couldn't get out of the car. She got stuck in the door and couldn't move in or out. . . ."

Her voice trailed off in helpless laughter. Everyone joined her: the men in mild deep voices, the women cackling high and shrill.

"Mercy me. How'd she ever get in in the first place?" Lil asked.

"I don't know," Audrey sobbed.

"What's more interesting," Henry said low beneath the women's laughter, "is how she got out."

It was a minute before Audrey could answer.

"We had to send one of the children, around through the other door . . . to push her. . . ."

"When is the last time anybody's been out to see Uncle Tom?" Sally asked woefully.

Audrey sobered.

"Aunt Sally, I don't reckon I've seen him since Anna's funeral."

"We haven't either," Lil said. "We ought to get together some Sunday and drive out to see him."

"We ought to," Lois agreed. "He'd like that."

Lois kissed Jefferson and his bride when they arrived and took them straight to her bedroom to leave their coats. Jefferson was nearest to her age and her favorite. She did not hold it against him, as some of the others did, because Mama had promised him a thousand dollars for his twenty-first birthday and he had gotten it from the estate after she died.

While he was still a boy he had come to Florida and spent the summer with her the year her second child was born. It had been the best time of her married life. They went fishing up the St. John River and her husband had been just as nice as he could be. She could talk to Jefferson; she was explaining to him about Bobby when Audrey came in to get her hat and coat.

"Audrey, where are you going?" Lois asked.

Outside an automobile horn sounded impatiently.

"That's Fred."

"Well, but what are you putting on your hat and coat for? Jefferson'll go out and help him bring in Fred junior."

"He's not coming in," Audrey said briskly. "We're going to his sister's house for dinner."

"Audrey Bronson!" Lois exclaimed. "Do you mean to tell me that after having brought all those cakes and having spent the whole morning here cooking . . . ?"

"I've spent half the day with my family, and I'll be fair and spend half the day with his."

Her voice was short. She hurried into the hall. Lois and Jefferson followed.

"Audrey . . ."

The doorbell rang as Audrey opened the door and walked past her husband on the porch.

"Didn't you hear me honking?" he asked.

"I was getting my hat and coat."

She continued across the porch and down the walk toward the car. Fred remained at the door, speaking to Lois. She could barely face him, she was so angry, but she was not going to give him the satisfaction of knowing it.

"Lois, I want to ask you something."

His voice was serious.

"What is it, Fred?"

She did not intend to let him ruffle her.

"I wonder if you can tell me why the Indian on the buffalo nickel has such a sneer on his face?"

He had taken out a nickel and held it in his hand. She looked from it to the sneer on his face but could think of no answer. He flipped the nickel over and his sneer broke into a snicker.

"Because his nose is so near to the buffalo's ass."

"Fred Bronson!"

Touching her fingertips to the bannisters, Lois led Jefferson up the long staircase. At the head of the stairs they could see into the room which had been his until he married. It was bare; he had taken the furniture. The floor ran from the doorsill to the hearth in orderly emptiness. At the room where Bobby was in bed they looked in and found him asleep, so they walked quietly back along the hall to the stairs. Hannah and her daughter clattered upward toward them with Cousin Ella following.

"You shouldn't talk that way to your mother, Honey," Cousin Ella called.

"Lady Jane! Listen to me."

Lady Jane stopped at the head of the stairs and stared at her mother.

"When your friends come I want you to bring them in and introduce them to me. Do you hear?"

"Yes," Lady Jane answered in anguish.

She ran down the upper hall to her room and slammed the door. Lois feared that Bobby would wake.

"Do what your mother tells you," Cousin Ella called after her.

"What's wrong?" Lois asked.

"She doesn't want to bring her friends in and introduce them to me before she goes out this afternoon," Hannah said. "But I'm not having her go out with any little buttermilks before I'm introduced to them."

She turned and started down the stairs. The others followed.

"She doesn't want them to see Lil and Sally using snuff," Cousin Ella whispered loudly to Lois.

Hannah stopped and turned. They stood strewn out along the stairs.

"Please, Cousin Ella, don't let Aunt Lil or Sally know anything about this."

"Oh, I wouldn't for the world," Cousin Ella protested. "I wouldn't say a word to anyone."

They clattered on down the stairs from vacancy to plenitude.

The sliding doors between the parlor and library were pushed open and the fire lit in the library. Besides more of the inlaid chairs and sofas of the other room, the library contained a large glass-doored bookcase flat against the wall, in which were kept a set of the Book of Knowledge and a set of law books as well as a typewritten bound history of the family of Hannah's husband. Both rooms were crowded now. The air curled with smoke and vibrated with voices loud and faint. The announcement of Audrey's departure broke in a wave of indignant sound which continued to agitate the merriment until Cy arrived. Hannah went out to his car with him to get the whisky for the eggnog and when she returned she whispered to Lois that she had told Cy they wanted to speak to him and he had agreed. It seemed that he wanted to speak to them, too. Lois did not quite hear what Hannah whispered, for names were bouncing from wall to wall of the room, rising higher and higher, but she gathered Hannah's meaning from her expression of knowing surprise.

When the sisters announced that dinner was ready, the entire assembly trooped back to the dining room carrying the air of consanguinity with them. The circle closed about the table. All the grownups except Hannah and Lois ate at first serving; they waited on table. At first there was one empty chair, for Cousin Ella wanted to help them; but at their insistence she finally sat down, asserting that it was only to keep the others company as she was not the least hungry. She did accept a cup of coffee beside her empty plate. Everyone except Hannah, who had learned from her husband to prefer it afterwards, drank coffee with the meal. And everyone continued to talk as the food was passed, so that only occasionally was a whole sentence by one person audible. Cousin Ella consented to take a little of the dark turkey and a buttered roll, just to keep up her constitution. Ivy's Negro and Mamie stayed in the kitchen and dished up the food while the sisters passed the roast and ham, the green vegetables and potatoes, the yams with marshmallow topping, then more turkey and dressing, the rolls, the condiments and jellies. They circled the table, complacently stupefied by the richness of the food before they had eaten any. Their movements and voices became infantile. Commanding and begging, they drawled back into the secure past as they sighed and lamented at the smallness of the portions taken. Cousin Ella refused more turkey but took a little dressing. When Henry's wife suggested that someone should be sent to wake Bobby, as it was a shame for anyone to miss such a feast, the sisters exchanged glances, for they did not want someone like her to know; but they felt no apprehension. Perhaps Ivy had talked. They smiled benignly and suggested that he might not feel like eating. Henry's wife said that she guessed maybe he didn't, and laughed raucously. The sisters dreamily moved on, serving in opposite directions. Cousin Ella took a little ham to go with her second cup of coffee. The meal became the happiest they remembered since Mama's death. Someone said that it was going to snow, and they all looked out the window past the empty dog yard which was perilously near to being overrun by the lightning vine. It was certainly dark, but the sisters did not think that it would snow. When they cleared the table Cousin Ella helped them, tossing into her mouth continual bits of bread and meat against the insatiable hunger which she feared might someday be unstaved. Hannah served the ambrosia. Lois put out the mincemeat pie, the chocolate cake, devil's food cake, coconut cake and walnut cake, the fruits and nuts. Aunt Sally declared that she must have gained ten pounds, and popped a glazed cherry into her mouth.

Rapidly, intent on dessert, at the second table the children ate. When they had finished, Hannah and Lois cleared the dining-room table and set the Negroes to washing dishes. Every surface in the kitchen was stacked with dirty plates and silver, and the sisters began to put the ingredients for eggnog, the platters, bowls and wire whisks, out on the table in the dining room. The late afternoon was falling and they turned on the chandelier. Cy and his wife appeared in the doorway to say that Audrey had returned and that they would have to leave soon.

The talk about Bobby no longer seemed as urgent to the sisters.

"We've got all the things out to make eggnog now," Hannah said.

"We can talk while you're making it."

"If you like," Hannah agreed.

Cy's wife stayed with Hannah and Lois while they waited for Cy to return with the family. As the wife of the only successful one of the brothers, she talked to Hannah and Lois as though she thought she was as superior to them as they felt they were to her, and the conversation was punctuated by a great deal of polite laughter. Cy's wife said that she envied them the wonderful eggnog they made and Hannah replied that in that case she could watch and see how they made it. The click of fork and whirl of eggbeater filled a short silence. Then Cy's wife said that she had never seen so large a sugar can. Lois laughed. It was small to them. When she had first married it had seemed so silly to go to the grocery store and come home with a little brown bag of sugar or a few slices of bacon wrapped in paper. Everything had been bought in barrels when they were girls: barrels of sugar, barrels of flour, barrels of meal, barrels of apples, everything—and whole sides of bacon.

Cy's wife waited until the whole family was in the dining room, just as though she were one of them, then left with an air of already knowing what was to be said. Hannah and Lois pursed their lips at each other but could not say anything with everyone there. Cy sat at the head of the table again. Hannah's fork continued to click and Lois's eggbeater to whine.

Cy began by saying that he did not know what they wanted to talk to him about, but that as he had to leave soon he would make his own say brief.

"We can make our say brief, too," Hannah said. Her wrist did not stop beating.

Henry said: "I didn't know we had anything to say. I thought Cy just wanted to talk to us."

Cy glanced toward Hannah. Without looking up from her work, she said:

"We sisters agreed to speak to Cy about Bobby."

"Something has to be done about him," Lois backed her up; "and if no one person is willing to help him the estate will have to."

"Help him how?" Henry asked.

"By advancing him enough money to start a filling station, or something."

"It looks like you just have to stay drunk to get what you want in this family."

"Henry, how can you say that?" Lois objected, not allowing herself to think that he had tried it enough to know.

"Think of Bobby's children," Hannah added and put down her fork.

Henry looked around at his brothers for support, but they did not like discussions. They preferred to take what was given them and to ask no questions. Then, if things went wrong, they could complain freely.

Triumphantly, Hannah dumped the yellow mixture into the cutglass punchbowl.

"Well, I see what Henry means," Ivy whined. "Bobby's not the only one. I've been sick for weeks and I've got all these doctor bills to pay."

"Well, you're not the only one, either," Hannah said, stirring in the whisky. "All of us have our troubles."

"I didn't say I was the only one," Ivy corrected her. "I said that Bobby wasn't."

"Let's not digress," Lois said. "Let's talk about Bobby like we agreed."

Ivy, huddled malignantly in her chair, twisted her head to face Lois. "Oh, you and Hannah can afford not to *digress*. You live here in the homeplace and it doesn't cost you a cent. That makes a lot of difference, don't you forget that."

Lois had never thought Ivy would turn on her like that. She replied with tears in her voice:

"Well, I certainly never thought you resented us living here. All the rest of you can live here. The only reason you don't is that you've got better places. And if we move out you'll only have to pay a caretaker."

"No one wants you to move out," Ivy said. "I'm just stating facts."

Jefferson reminded them that they were to talk about Bobby. Lois would not have been surprised then if someone had made a remark about favorites, but no one did. Hannah folded the egg whites into the bowl. Cy stood up. Left with nothing to do, Lois held out the nutmeg which Hannah did not yet need.

"Well, I'm here to tell you that there's less money, not more," Cy said, "so if whatever you want to do for Bobby takes money, you'd better forget it."

Audrey had sat in her seat with her lips pressed defiantly together, but now she parted them.

"Why?" she snapped. "Is it your money? I'd like to know why."

Ivy moaned to indicate that the whole discussion was too much for an invalid. But Cy was angry.

"Any time anyone wants to take charge of the estate I'll be glad to put the whole damned mess into his hands," he said.

"Mess? How did it get into a mess?" Audrey asked quickly.

"I worry about the estate all year around. Then, once a year, you all get excited about money, money, money, and start complaining. Well, you can take charge of it. All of it."

Lois tried to deny that she had ever complained, but Audrey was leaning across the table shaking her finger at Cy.

"You needn't be so high and mighty just because you have a lot coming in and the rest of us only have it going out," she cried. "Just let me remind you that if it hadn't been for your brothers and sisters working for you up to midnight on the very eve of election you wouldn't have that job of yours. And maybe if everything was above-board and honest you wouldn't still have it anyway. It's mighty funny to me that before the next election you made your opponent that you'd said all those awful things about into your assistant and the position was changed to a permanent one. Everybody doesn't have such good *luck*."

"Audrey!" Lois exclaimed. "That's your own brother you're talking to. Leave politics to politicians."

Cy grew red in the face and slammed his fist to the table.

"If you're suggesting that I've done anything illegal with Mama's estate I can show you papers for every transaction I've made. And I can give you an account of my own time and money spent."

"Cy," Hannah pleaded, "Audrey didn't mean what she said."

"Yes, I did," Audrey cried. "I'd like to see his accounts. If we all need money, I'd like to know why we can't all have it. It's Mama's money, not Cy's. It's ours."

"Mama would turn over in her grave if she heard us talking like this to each other," Lois said.

"Sit down and let Cy talk," Stuart said to Audrey.

"Let him talk," Audrey said. "Let him talk. I'm ready to listen."

Cy stood for a long time, silent, staring at Audrey who had no intention of being outstared. At last he looked around the circle at the table and spoke shortly.

"What I was going to say before Audrey's outburst was that Mama's securities have dropped in value. Lawyer Smith made some bad investments. You have either to take lower incomes or to sell a piece of property. All the property that's left just eats up money in taxes, anyway."

"What about our income from the Grand Hotel? What about that?" Audrey demanded. "We still own half of that, don't we?"

"That place isn't worth anything except as land, and you can't sell unless Lawyer Smith agrees to sell his half."

"I don't know anything about business," Audrey said, "but it's mighty funny to me that there isn't any income from a hotel that's doing business all the time."

"You don't know anything about what kind of place that hotel is! It's not even on the street. You have to go down an alley to reach it."

"Let's hear what Cy has to say," Hannah told Audrey.

There was another silence.

"Let Audrey say. It's not up to me."

In the cutglass bowl, the eggnog, unstirred, separated, yellow to the bottom, cream to the top.

At last Cy repeated that Mama's securities had dropped in value until there was no income beyond that which was necessary to pay the yearly taxes on her property. If the brothers and sisters were to continue receiving money, property would have to be sold. The best piece to begin with would be the vacant lot over on Georgia Avenue. It was not worth much, but it never would be worth much; and it would bring in enough to pay this year's taxes and give each of them a small sum. But not a sum large enough to set anyone up in business. If they wanted to do that there was only one piece of property valuable enough. The homeplace.

"I don't believe a word you're saying," Audrey said. "You just say that because you know none of us want to sell the homeplace."

"Then I won't say any more."

Cy went out through the door to the hall and closed it softly behind him. Lois caught up with him on the front porch just as he was leaving the house. Outside in the world, she saw his wife already waiting for him in their car. Cy did not stop when she called his name. She ran down the steps and along the herringbone brick wall which divided the yard and placed her hand on his arm. He stopped and turned to face her but he did not say anything.

"Cy, you're not going to sell the homeplace, are you?"

"I think it would be a very foolish thing to do until the right time," he replied.

For a while everyone was separated, like children after a quarrel. Standing in the bay of windows beside her Christmas tree, Lois looked out at the unpierced night sky. Snow had begun to fall. Slowly, the ornaments of the boardinghouse across the street were outlined in white and the house itself disappeared against the black sky. And suddenly she was crying without thinking why, crying as though the cause of her grief lay in the cold window glass and bare yard, the water faucet wrapped in crocus sacking, and the buildings disappearing in the snow. Toward the snow of falling cornices and fading spindles she felt the fear which she had felt toward strange places and people not her kind. In the white disappearing of gables and mansards, the walls of the homeplace dissolved and left her bare to the world. All the things which she loved had come from that world: her children; her diamond rings and pearshaped buttons; the ornaments on the tree at her side, saved year after year since the first Christmas of her marriage. But she had thought that she would never leave the homeplace again, that she and her children would stay here always with the many people death had undone in the house, and she did not understand what had happened.

The room was dark except for the fire and the ornamental lights. She turned toward the door to switch on the chandelier. As she passed the tree her arm brushed against a metal bird with spring legs and a spun glass tail, and as she crossed the room the bird wobbled up and down, this way and that.

"Mother?"

She stopped beside the couch and kissed her son's forehead.

"Yes, darling."

"Where are you going?"

"Nowhere, darling. Mother's going to stay right here with you."

"What's the matter?"

"Nothing, darling. Mother just wants you to get well. Hush now and go back to sleep."

From the silence of the house, Hannah's footsteps echoed. The door opened and a din of voices, exclamations, cries of disbelief and laughter came in a tumult of sound across the hall.

"Wasn't that scene a shame!" Hannah said, striking her hands together. "Audrey can't learn to hold her temper. But she's forgotten it already. She's in there talking and laughing with Aunt Lil and Sally as though nothing has happened."

Her face bereft of expression, Lois looked up at her sister.

"Hannah, what are we going to do if Cy sells the homeplace?"

"There's no chance of his doing that anytime soon," Hannah said.

"But, Hannah, what are we going to do?"

"Well," Hannah said, "I'm going back now and dish up the eggnog. As soon as you think I've had time to cut the fruitcake, tell them all to come on back."

Women Lost

Lost ladies, dropped out of the bottom of life, usually by loneliness or age, by unreturned love, by the accumulated pressure of situations they cannot handle—what a range of great women in literature can be found here! Anna Karenina and Emma Bovary for a start, and in the theater Phaedra and Ophelia, Miss Julie, Hedda Gabler and Rebecca West, Yerma and Blanche Dubois! Marian Forrester. The heroines of Jean Rhys, Anna in Voyages in the Dark, Sasha in Good Morning, Midnight, the mad Creole heiress in Wide Sargasso Sea. Judith Hearne taking to the bottle in the icy loneliness of her pathetic life in Belfast. Peyton Loftis trapped in Virginia. Again, the difficulty was in choosing. I decided on one exceptional poem and three descriptions of types excerpted from novels of classic stature.

Hetty Sorel in Adam Bede is a beautiful young girl who loses her head to a man above her in the social scale, is seduced, has her baby secretly, kills it for fear of detection, is discovered, tried and almost hanged but finally sent to Australia, there to expiate her sin by a life of repentant drudgery. Carbon copies of Hetty are the stuff of cheap romantic fiction. Arthur Donnithorne, the seducer, also must pay for his transgression (for Eliot is an author of the most impeccable morality), but of course his life is not quite ruined the way poor Hetty's is.

In this chapter George Eliot analyzes Hetty's charm and vanity with insight into her character and also with a little of that combination of envy and contempt that an ugly and brilliant woman would feel for a beautiful and foolish one. Hetty is contrasted in the novel with Dinah Morris, the itinerant Methodist lay preacher who exudes an air of quiet serenity and religious joy and ends up as Adam Bede's wife after Hetty's fall. Hetty is the more vivid character, however, a triumph of *observation and understanding on Eliot's part, a young girl whose type is by no means obsolete, even though today her history and fate might well be a bit different.*

Dickens is thought of as a writer unable to create convincing women characters, and in general that is true if we consider his young girls, generally cardboard characters of repellent sweetness—although Estella in Great Expectations and Bella Wilfer in Our Mutual Friend are complex and reasonably convincing young women. He does much better with his comic figures like Sairey Gamp, Mrs. Micawber or Betsey Trotwood, women safely older or well married.

The chapter from Little Dorrit which I have included is unique in his work, being a realistic analysis of Miss Wade, a minor character in a thickly populated novel. "The History of a Self Tormentor" is self-contained and plays no role in the complicated plot; it must have come out of some suddenly compelling need in Dickens himself.

Despite his gift for caricature, Dickens was a meticulous observer, and physicians are able to diagnose the illnesses that afflicted his characters from his descriptions of symptoms. The same is true with the psychiatric observation of Miss Wade, an intensely neurotic and isolated woman who suffers bitterly from her recognition of the lowly role women must play in her society, particularly if they are neither beautiful nor rich. Her life is poisoned by unacknowledged erotic feelings, shut up inside a cold exterior and a flinty intelligence that keep her from believing in any evidence of sympathy or affection she may receive.

Nightwood is a novel of a lost world, peopled by the lost, the main ones being the lovers Nora Flood and Robin Vote, wealthy Nora sheltering Robin, who has fled from her husband and child and who stays with Nora only for a time, then continues her curiously innocent journey to destruction. In the "Night Watch" chapter the paths of the two women intersect for a brief moment of equilibrium. The contrast between the decadence and sensuality of the subject and the controlled glitter of the style gives the book a hypnotic power unmatched in American literature.

Anne Sexton is a poet of extreme states who writes about madness, desperate love, and the loss of self. In "Consorting with Angels" she celebrates an alienation that ends in triumph.

ANNE SEXTON
(1 9 2 8 –   )

Consorting with Angels

I was tired of being a woman,
tired of the spoons and the pots,
tired of my mouth and my breasts,
tired of the cosmetics and the silks.
There were still men who sat at my table,
circled around the bowl I offered up.
The bowl was filled with purple grapes
and the flies hovered in for the scent
and even my father came with his white bone.
But I was tired of the gender of things.

Last night I had a dream
and I said to it . . .
"You are the answer.
You will outlive my husband and my father."
In that dream there was a city made of chains
where Joan was put to death in man's clothes
and the nature of the angels went unexplained,
no two made in the same species,
one with a nose, one with an ear in its hand,
one chewing a star and recording its orbit,
each one like a poem obeying itself,
performing God's functions,
a people apart.

"You are the answer,"
I said, and entered,
lying down on the gates of the city.
Then the chains were fastened around me
and I lost my common gender and my final aspect.
Adam was on the left of me
and Eve was on the right of me,
both thoroughly inconsistent with the world of reason.
We wove our arms together
and rode under the sun.
I was not a woman anymore,
not one thing or the other.

O daughters of Jerusalem,
the king has brought me into his chamber.
I am black and I am beautiful.
I've been opened and undressed.
I have no arms or legs.
I'm all one skin like a fish.
I'm no more a woman
than Christ was a man.

GEORGE ELIOT
(MARY ANN EVANS, 1819–1880)

Adam Bede

Hetty and Dinah both slept in the second story, in rooms adjoining each other, meagrely-furnished rooms, with no blinds to shut out the light, which was now beginning to gather new strength from the rising of the moon—more than enough strength to enable Hetty to move about and undress with perfect comfort. She could see quite well the pegs in the old painted linen-press on which she hung her hat and gown; she could see the head of every pin on her red cloth pin-cushion; she could see a reflection of herself in the old-fashioned looking-glass, quite as distinct as was needful, considering that she had only to brush her hair and put on her night-cap. A queer old looking-glass! Hetty got into an ill-temper with it almost every time she dressed. It had been considered a handsome glass in its day, and had probably been bought into the Poyser family a quarter of a century before, at a sale of genteel household furniture. Even now an auctioneer could say something for it: it had a great deal of tarnished gilding about it; it had a firm mahogany base, well supplied with drawers, which opened with a decided jerk, and sent the contents leaping out from the farthest corners, without giving you the trouble of reaching them; above all, it had a brass candle-socket on each side, which would give it an aristocratic air to the very last. But Hetty objected to it because it had numerous dim blotches sprinkled over the mirror, which no rubbing would remove, and because, instead of swinging backwards and forwards, it was fixed in an upright position, so that she could only get one good view of her head and neck, and that was to be had only by sitting down on a low chair before her dressing-table. And the dressing-table was no dressing-table at all, but a small old chest of drawers, the most awkward thing in the world to sit down before, for the big brass handles quite hurt her knees, and she couldn't get near the glass at all comfortably. But devout worshippers never allow inconveniences to prevent them from performing their religious rites, and Hetty this evening was more bent on her peculiar form of worship than usual.

Having taken off her gown and white kerchief, she drew a key from the large pocket that hung outside her petticoat, and, unlocking one of the lower drawers in the chest, reached from it two short bits of wax candle—secretly bought at Treddleston—and stuck them in the two brass sockets. Then she drew forth a bundle of matches, and lighted the candles; and last of all, a small red-framed shilling looking-glass, without blotches. It was into this small glass that she chose to look first after seating herself. She looked into it, smiling, and turning her head on one side, for a minute, then laid it down and took out her brush and comb from an upper drawer. She was going to let down her hair, and make herself look like that picture of a lady in Miss Lydia Donnithorne's dressing-room. It was soon done, and the dark hyacinthine curves fell on her neck. It was not heavy, massive, merely rippling hair, but soft and silken, running at every opportunity into delicate rings. But she pushed it all backward to look like the picture, and form a dark curtain, throwing into relief her round white neck. Then she put down her brush and comb, and looked at herself, folding her arms before her, still like the picture. Even the old mottled glass couldn't help sending back a lovely image, none the less lovely because Hetty's stays were not of white satin—such as I feel sure heroines must generally wear—but of a dark greenish cotton texture.

O yes! she was very pretty: Captain Donnithorne thought so. Prettier than anybody about Hayslope—prettier than any of the ladies she had ever seen visiting at the Chase—indeed it seemed fine ladies were rather old and ugly—and prettier than Miss Bacon, the miller's daughter, who was called the beauty of Treddleston. And Hetty looked at herself to-night with quite a different sensation from what she had ever felt before; there was an invisible spectator whose eye rested on her like morning on the flowers. His soft voice was saying over and over again those pretty things she had heard in the wood; his arm was round her, and the delicate rose-scent of his hair was with her still. The vainest woman is never thoroughly conscious of her own beauty till she is loved by the man who sets her own passion vibrating in return.

But Hetty seemed to have made up her mind that something was wanting, for she got up and reached an old black lace scarf out of the linen-press, and a pair of large earrings out of the sacred drawer from which she had taken her candles. It was an old old scarf, full of rents, but it would make a becoming border round her shoulders, and set off the whiteness of her upper arm. And she would take out the little earrings she had in her ears—oh, how her aunt had scolded her for having her ears bored!—and put in those large ones: they were but coloured glass and gilding; but if you didn't know what they were made of, they looked just as well as what the ladies wore. And so she sat down again, with the large earrings in her ears, and the black lace scarf adjusted round her shoulders. She looked down at her arms: no arms could be prettier down to a little way below the elbow—they were white and plump, and dimpled to match her cheeks; but towards the wrist, she thought with vexation that they were coarsened by butter-making, and other work that ladies never did.

Captain Donnithorne couldn't like her to go on doing work: he would like to see her in nice clothes, and thin shoes and white stock-

ings, perhaps with silk clocks to them; for he must love her very much
—no one else had ever put his arm round her and kissed her in that
way. He would want to marry her, and make a lady of her; she could
hardly dare to shape the thought—yet how else could it be? Marry her
quite secretly, as Mr. James, the Doctor's assistant, married the Doctor's
niece, and nobody ever found it out for a long while after, and then it
was of no use to be angry. The Doctor had told her aunt all about it
in Hetty's hearing. She didn't know how it would be, but it was quite
plain the old Squire could never be told anything about it, for Hetty
was ready to faint with awe and fright if she came across him at the
Chase. He might have been earth-born, for what she knew: it had
never entered her mind that he had been young like other men; he
had always been the old Squire at whom everybody was frightened.
O it was impossible to think how it would be! But Captain Donnithorne would know; he was a great gentleman, and could have his
way in everything, and could buy everything he liked. And nothing
could be as it had been again: perhaps some day she should be a
grand lady, and ride in her coach, and dress for dinner in a brocaded
silk, with feathers in her hair, and her dress sweeping the ground, like
Miss Lydia and Lady Dacey, when she saw them going into the dining-
room one evening, as she peeped through the little round window in
the lobby; only she should not be old and ugly like Miss Lydia, or all
the same thickness like Lady Dacey, but very pretty, with her hair done
in a great many different ways, and sometimes in a pink dress, and
sometimes in a white one—she didn't know which she liked best; and
Mary Burge and everybody would perhaps see her going out in her
carriage—or rather, they would *hear* of it: it was impossible to imagine
these things happening at Hayslope in sight of her aunt. At the
thought of all this splendour, Hetty got up from her chair, and in
doing so caught the little red-framed glass with the edge of her scarf,
so that it fell with a bang on the floor; but she was too eagerly occupied
with her vision to care about picking it up; and after a momentary
start, began to pace with a pigeon-like stateliness backwards and forwards along her room, in her coloured stays and coloured skirt, and the
old black lace scarf round her shoulders, and the great glass earrings in
her ears.

How pretty the little puss looks in that odd dress! It would be the easiest folly in the world to fall in love with her: there is such a sweet baby-like roundness about her face and figure; the delicate dark rings of hair lie so charmingly about her ears and neck; her great dark eyes with their long eyelashes touch one so strangely, as if an imprisoned frisky sprite looked out of them.

Ah, what a prize the man gets who wins a sweet bride like Hetty! How the men envy him who come to the wedding breakfast, and see her hanging on his arm in her white lace and orange blossoms. The dear, young, round, soft, flexible thing! Her heart must be just as soft, her temper just as free from angles, her character just as pliant. If anything ever goes wrong, it must be the husband's fault there: he can make her what he likes—that is plain. And the lover himself thinks so too: the little darling is so fond of him, her little vanities are so bewitching, he wouldn't consent to her being a bit wiser; those kitten-like glances and movements are just what one wants to make one's heart a paradise. Every man under such circumstances is conscious of being a great physiognomist. Nature, he knows, has a language of her own, which she uses with strict veracity, and he considers himself an adept in the language. Nature has written out his bride's character for him in those exquisite lines of cheek and lip and chin, in those eyelids delicate as petals, in those long lashes curled like the stamen of a flower, in the dark liquid depths of those wonderful eyes. How she will dote on her children! She is almost a child herself, and the little pink round things will hang about her like florets round the central flower; and the husband will look on, smiling benignly, able, whenever he chooses, to withdraw into the sanctuary of his wisdom, towards which his sweet wife will look reverently, and never lift the curtain. It is a marriage such as they made in the golden age, when the men were all wise and majestic, and the women all lovely and loving.

It was very much in this way that our friend Adam Bede thought about Hetty; only he put his thoughts into different words. If ever she behaved with cold vanity towards him, he said to himself, it is only because she doesn't love me well enough; and he was sure that her love, whenever she gave it, would be the most precious thing a man could possess on earth. Before you despise Adam as deficient in penetration, pray ask yourself if you were ever predisposed to believe evil of any pretty woman—if you ever *could,* without hard head-breaking demonstration, believe evil of the *one* supremely pretty woman who has be-witched you. No: people who love downy peaches are apt not to think of the stone, and sometimes jar their teeth terribly against it.

Arthur Donnithorne, too, had the same sort of notion about Hetty, so far as he had thought of her nature at all. He felt sure she was a dear, affectionate, good little thing. The man who awakes the wonder-ing tremulous passion of a young girl always thinks her affectionate; and if he chances to look forward to future years, probably imagines himself being virtuously tender to her, because the poor thing is so clingingly fond of him. God made these dear women so—and it is a convenient arrangement in case of sickness.

After all, I believe the wisest of us must be beguiled in this way sometimes, and must think both better and worse of people than they deserve. Nature has her language, and she is not unveracious; but we don't know all the intricacies of her syntax just yet, and in a hasty reading we may happen to extract the very opposite of her real mean-ing. Long dark eyelashes, now: what can be more exquisite? I find it impossible not to expect some depth of soul behind a deep grey eye with a long dark eyelash, in spite of an experience which has shown me that they may go along with deceit, peculation, and stupidity. But if, in the reaction of disgust, I have betaken myself to a fishy eye, there has been a surprising similarity of result. One begins to suspect at length that there is no direct correlation between eyelashes and morals; or else, that the eyelashes express the disposition of the fair one's grand-mother, which is on the whole less important to us.

No eyelashes could be more beautiful than Hetty's; and now, while she walks with her pigeon-like stateliness along the room and looks down on her shoulders bordered by the old black lace, the dark fringe shows to perfection on her pink cheek. They are but dim ill-defined pictures that her narrow bit of an imagination can make of the future; but of every picture she is the central figure in fine clothes; Captain Donnithorne is very close to her, putting his arm round her, perhaps kissing her, and everybody else is admiring and envying her—especially Mary Burge, whose new print dress looks very contemptible by the side of Hetty's resplendent toilette. Does any sweet or sad memory mingle with this dream of the future—any loving thought of her sec-ond parents—of the children she had helped to tend—of any youthful companion, any pet animal, any relic of her own childhood even? Not one. There are some plants that have hardly any roots: you may tear them from their native nook of rock or wall, and just lay them over your ornamental flower-pot, and they blossom none the worse. Hetty could have cast all her past life behind her, and never cared to be reminded of it again. I think she had no feeling at all towards the old house, and did not like the Jacob's Ladder and the long row of holly-hocks in the garden better than other flowers—perhaps not so well. It was wonderful how little she seemed to care about waiting on her uncle, who had been a good father to her: she hardly ever remembered to reach him his pipe at the right time without being told, unless a visitor happened to be there, who would have a better opportunity of seeing her as she walked across the hearth. Hetty did not understand how anybody could be very fond of middle-aged people. And as for those tiresome children, Marty and Tommy and Totty, they had been the very nuisance of her life—as bad as buzzing insects that will come teasing you on a hot day when you want to be quiet. Marty, the eldest, was a baby when she first came to the farm, for the children born before him had died, and so Hetty had had them all three, one after the other, toddling by her side in the meadow, or playing about her on wet days in the half-empty rooms of the large old house. The boys were out of hand now, but Totty was still a day-long plague, worse than either of the others had been, because there was more fuss made about her. And there was no end to the making and mending of clothes. Hetty would have been glad to hear that she should never see a child again; they were worse than the nasty little lambs that the shepherd was always bringing in to be taken special care of in lambing time; for the lambs *were* got rid of sooner or later. As for the young chickens and turkeys, Hetty would have hated the very word "hatching," if her aunt had not bribed her to attend to the young poultry by promising her the proceeds of one out of every brood. The round downy chicks peeping out from under their mother's wing never touched Hetty with any pleasure; that was not the sort of prettiness she cared about, but she did care about the prettiness of the new things she would buy for herself at Treddleston fair with the money they fetched. And yet she looked so dimpled, so charming, as she stooped down to put the soaked bread under the hencoop, that you must have been a very acute personage indeed to suspect her of that hardness. Molly, the housemaid, with a turn-up nose and a protuberant jaw, was really a tender-hearted girl, and, as Mrs. Poyser said, a jewel to look after the poultry; but her stolid face showed nothing of this maternal delight, any more than a brown earthenware pitcher will show the light of the lamp within it.

It is generally a feminine eye that first detects the moral deficiencies hidden under the "dear deceit" of beauty: so it is not surprising that Mrs. Poyser, with her keenness and abundant opportunity for observation, should have formed a tolerably fair estimate of what might be expected from Hetty in the way of feeling, and in moments of indignation she had sometimes spoken with great openness on the subject to her husband.

"She's no better than a peacock, as 'ud strut about on the wall, and spread its tail when the sun shone if all the folks i' the parish was dying: there's nothing seems to give her a turn i' th' inside, not even when we thought Totty had tumbled into the pit. To think o' that dear cherub! And we found her wi' her little shoes stuck i' the mud an' crying fit to break her heart by the far horse-pit. But Hetty niver minded it, I could see, though she's been at the nussin' o' the child ever since it was a babby. It's my belief her heart's as hard as a pebble."

"Nay, nay," said Mr. Poyser, "thee mustn't judge Hetty too hard. Them young gells are like th' unripe grain; they'll make good meal by-and-by, but they're squashy as yet. Thee't see Hetty'll be all right when she's got a good husband and children of her own."

"*I* don't want to be hard upo' the gell. She's got cliver fingers of her own, and can be useful enough when she likes, and I should miss her wi' the butter, for she's got a cool hand. An' let be what may, I'd strive to do my part by a niece o' yours, an' *that* I've done: for I've taught her everything as belongs to a house, an' I've told her her duty often enough, though, God knows, I've no breath to spare, an' that catchin' pain comes on dreadful by times. Wi' them three gells in the house I'd need have twice the strength, to keep 'em up to their work. It's like having roast-meat at three fires; as soon as you've basted one, another's burnin'."

Hetty stood sufficiently in awe of her aunt to be anxious to conceal from her so much of her vanity as could be hidden without too great a sacrifice. She could not resist spending her money in bits of finery which Mrs. Poyser disapproved; but she would have been ready to die with shame, vexation, and fright, if her aunt had this moment opened the door, and seen her with her bits of candle lighted, and strutting about decked in her scarf and earrings. To prevent such a surprise, she always bolted her door, and she had not forgotten to do so to-night. It was well: for there now came a light tap, and Hetty with a leaping heart, rushed to blow out the candles and throw them into the drawer. She dared not stay to take out her earrings, but she threw off her scarf, and let it fall on the floor, before the light tap came again. We shall know how it was that the light tap came, if we leave Hetty for a short time, and return to Dinah, at the moment when she had delivered Totty to her mother's arms, and was come up-stairs to her bedroom, adjoining Hetty's.

CHARLES DICKENS
(1812–1870)

Little Dorrit

I have the misfortune of not being a fool. From a very early age I have detected what those about me thought they hid from me. If I could have been habitually imposed upon, instead of habitually discerning the truth, I might have lived as smoothly as most fools do.

My childhood was passed with a grandmother; that is to say, with a lady who represented that relative to me; and who took that title on herself. She had no claim to it, but I—being to that extent a little fool —had no suspicion of her. She had some children of her own family in her house, and some children of other people. All girls; ten in number, including me. We all lived together and were educated together.

I must have been about twelve years old when I began to see how determinedly those girls patronised me. I was told I was an orphan. There was no other orphan among us; and I perceived (here was the first advantage of not being a fool) that they conciliated me in an insolent pity, and in a sense of superiority. I did not set this down as a discovery, rashly. I tried them often. I could hardly make them quarrel with me. When I succeeded with any of them, they were sure to come after an hour or two, and begin a reconciliation. I tried them over and over again, and I never knew them wait for me to begin. They were always forgiving me, in their vanity and condescension. Little images of grown people!

One of them was my chosen friend. I loved that stupid mite in a passionate way that she could no more deserve, than I can remember without feeling ashamed of, though I was but a child. She had what they call an amiable temper, an affectionate temper. She could distribute, and did distribute, pretty looks and smiles to every one among them. I believe there was not a soul in the place, except myself, who knew that she did it purposely to wound and gall me!

Nevertheless, I so loved that unworthy girl, that my life was made stormy by my fondness for her. I was constantly lectured and disgraced for what was called "trying her"; in other words, charging her with her little perfidy and throwing her into tears by showing her that I read her heart. However, I loved her, faithfully; and one time I went home with her for the holidays.

She was worse at home than she had been at school. She had a crowd of cousins and acquaintances, and we had dances at her house, and went out to dances at other houses. Both at home and out, she tormented my love beyond endurance. Her plan was, to make them all fond of her—and so drive me wild with jealousy. To be familiar and endearing with them all—and so make me mad with envying them. When we were left alone in our bedroom at night, I would reproach her with my perfect knowledge of her baseness; and then she would cry and cry and say I was cruel, and then I would hold her in my arms till morning: loving her as much as ever, and often feeling as if, rather than suffer so, I could so hold her in my arms and plunge to the bottom of a river—where I would still hold her, after we were both dead.

It came to an end, and I was relieved. In the family, there was an aunt, who was not fond of me. I doubt if any of the family liked me much; but, I never wanted them to like me, being altogether bound up in the one girl. The aunt was a young woman, and she had a serious way with her eyes of watching me. She was an audacious woman, and openly looked compassionately at me. After one of the nights that I have spoken of, I came down into a greenhouse before breakfast. Charlotte (the name of my false young friend) had gone down before me, and I heard this aunt speaking to her about me as I entered. I stopped where I was, among the leaves, and listened.

The aunt said, "Charlotte, Miss Wade is wearing you to death, and this must not continue." I repeat the very words I heard.

Now, what did she answer? Did she say, "It is I who am wearing her to death, I who am keeping her on a rack and am the executioner, yet she tells me every night that she loves me devotedly, though she knows what I make her undergo?" No; my first memorable experience was true to what I knew her to be, and to all my experience. She began sobbing and weeping (to secure the aunt's sympathy to herself), and said, "Dear aunt, she has an unhappy temper; other girls at school, besides I, try hard to make it better; we all try hard."

Upon that, the aunt fondled her, as if she had said something noble instead of despicable and false, and kept up the infamous pretence by replying, "But there are reasonable limits, my dear love, to everything, and I see that this poor miserable girl causes you more constant and useless distress than even so good an effort justifies."

The poor miserable girl came out of her concealment, as you may be prepared to hear, and said, "Send me home." I never said another word to either of them, or to any of them, but "Send me home, or I will walk home alone, night and day!" When I got home, I told my supposed grandmother that, unless I was sent away to finish my education somewhere else, before that girl came back, or before any one of them came back, I would burn my sight away by throwing myself into the fire, rather than I would endure to look at their plotting faces.

I went among young women next, and I found them no better. Fair words and fair pretences; but, I penetrated below those assertions of themselves and depreciations of me, and they were no better. Before I left them, I learned that I had no grandmother and no recognised relation. I carried the light of that information both into my past and into my future. It showed me many new occasions on which people triumphed over me, when they made a pretence of treating me with consideration, or doing me a service.

A man of business had a small property in trust for me. I was to be a governess. I became a governess; and went into the family of a poor nobleman, where there were two daughters—little children, but the parents wished them to grow up, if possible, under one instructress. The mother was young and pretty. From the first, she made a show of behaving to me with great delicacy. I kept my resentment to myself; but, I knew very well that it was her way of petting the knowledge that she was my Mistress, and might have behaved differently to her servant if it had been her fancy.

I say I did not resent it, nor did I; but I showed her, by not gratifying her, that I understood her. When she pressed me to take wine I took water. If there happened to be anything choice at table, she always sent it to me: but I always declined it, and ate of the rejected dishes. These disappointments of her patronage were a sharp retort, and made me feel independent.

I liked the children. They were timid, but on the whole disposed to attach themselves to me. There was a nurse, however, in the house, a rosy-faced woman always making an obtrusive pretence of being gay and good-humoured, who had nursed them both, and who had secured their affections before I saw them. I could almost have settled down to my fate but for this woman. Her artful devices for keeping herself before the children in constant competition with me, might have blinded many in my place; but I saw through them from the first. On the pretext of arranging my rooms and waiting on me and taking care of my wardrobe (all of which she did busily), she was never absent. The most crafty of her many subtleties was her feint of seeking to make the children fonder of me. She would lead them to me and coax them to me. "Come to good Miss Wade, come to dear Miss Wade, come to pretty Miss Wade. She loves you very much. Miss Wade is a clever lady, who has read heaps of books and can tell you far better and more interesting stories than I know. Come and hear Miss Wade!" How could I engage their attention, when my heart was burning against these ignorant designs? How could I wonder when I saw their innocent faces shrinking away, and their arms twining round her neck instead of mine? Then she would look up at me, shaking their curls from her face, "They'll come round soon, Miss Wade; they're very simple and loving, ma'am; don't be at all cast down about it, ma'am"—exulting over me!

There was another thing the woman did. At times, when she saw that she had safely plunged me into a black despondent brooding by these means, she would call the attention of the children to it, and would show them the difference between herself and me. "Hush! Poor Miss Wade is not well. Don't make a noise, my dears, her head aches. Come and comfort her. Come and ask her if she is better; come and ask her to lie down. I hope you have nothing on your mind, ma'am. Don't take on ma'am, and be sorry!"

It became intolerable. Her ladyship my Mistress coming in one day when I was alone, and at the height of feeling that I could support it no longer, I told her I must go. I could not bear the presence of that woman Dawes.

"Miss Wade! Poor Dawes is devoted to you; would do anything for you!"

I knew beforehand she would say so; I was quite prepared for it; I only answered, it was not for me to contradict my Mistress; I must go.

"I hope, Miss Wade," she returned, instantly assuming the tone of superiority she had always so thinly concealed, "that nothing I have ever said or done since we have been together, has justified your use of that disagreeable word, Mistress. It must have been wholly inadvertent on my part. Pray tell me what it is."

I replied that I had no complaint to make, either of my Mistress or to my Mistress; but, I must go.

She hesitated a moment, and then sat down beside me, and laid her hand on mine. As if that honour would obliterate any remembrance!

"Miss Wade, I fear you are unhappy, through causes over which I have no influence."

I smiled, thinking of the experience the word awakened, and said, "I have an unhappy temper, I suppose."

"I did not say that."

"It is an easy way of accounting for anything," said I.

"It may be; but I did not say so. What I wish to approach, is something very different. My husband and I have exchanged remarks upon the subject, when we have observed with pain that you have not been easy with us."

"Easy? Oh! You are such great people, my lady," said I.

"I am unfortunate in using a word which may convey a meaning— and evidently does—quite opposite to my intention." (She had not expected my reply, and it shamed her.) "I only mean not happy with us. It is a difficult topic to enter on; but, from one young woman to another, perhaps—in short, we have been apprehensive that you may

allow some family circumstances of which no one can be more innocent than yourself, to prey upon your spirits. If so, let us entreat you
not to make them a cause of grief. My husband himself, as is well
known, formerly had a very dear sister who was not in law his sister,
but who was universally beloved and respected—"

I saw directly, that they had taken me in, for the sake of the dead
woman, whoever she was, and to have that boast of me and advantage
of me; I saw, in the nurse's knowledge of it, an encouragement to
goad me as she had done; and I saw, in the children's shrinking away,
a vague impression that I was not like other people. I left that house
that night.

After one or two short and very similar experiences, which are not
to the present purpose, I entered another family where I had but one
pupil: a girl of fifteen, who was the only daughter. The parents here
were elderly people: people of station and rich. A nephew whom they
had brought up, was a frequent visitor at the house, among many other
visitors; and he began to pay me attention. I was resolute in repulsing
him; for, I had determined when I went there, that no one should
pity me or condescend to me. But, he wrote me a letter. It led to our
being engaged to be married.

He was a year younger than I, and young-looking even when that
allowance was made. He was on absence from India, where he had a
post that was soon to grow into a very good one. In six months we were
to be married, and were to go to India. I was to stay in the house, and
was to be married from the house. Nobody objected to any part of the
plan.

I cannot avoid saying, he admired me; but, if I could, I would. Vanity
has nothing to do with the declaration, for, his admiration worried me.
He took no pains to hide it; and caused me to feel among the rich
people as if he had bought me for my looks, and made a show of his
purchase to justify himself. They appraised me in their own minds, I
saw, and were curious to ascertain what my full value was. I resolved
that they should not know. I was immovable and silent before them;
and would have suffered any one of them to kill me sooner than I
would have laid myself out to bespeak their approval.

He told me I did not do myself justice. I told him I did, and it
was because I did and meant to do so to the last, that I would not stoop
to propitiate any of them. He was concerned and even shocked, when

I added that I wished he would not parade his attachment before them; but, he said he would sacrifice even the honest impulses of his affection to my peace.

Under that pretence he began to retort upon me. By the hour together, he would keep at a distance from me, talking to any one rather than to me. I have sat alone and unnoticed, half an evening, while he conversed with his young cousin, my pupil. I have seen all the while, in people's eyes, that they thought the two looked nearer on an equality than he and I. I have sat, divining their thoughts, until I have felt that his young appearance made me ridiculous, and have raged against myself for ever loving him.

For, I did love him once. Undeserving as he was, and little as he thought of all these agonies that it cost me—agonies which should have made him wholly and gratefully mine to his life's end—I loved him. I bore with his cousin's praising him to my face, and with her pretending to think that it pleased me, but full well knowing that it rankled in my breast; for his sake. While I have sat in his presence recalling all my slights and wrongs, and deliberating whether I should not fly from the house at once and never see him again—I have loved him.

His aunt (my Mistress, you will please to remember) deliberately, wilfully, added to my trials and vexations. It was her delight to expatiate on the style in which we were to live in India, and on the establishment we should keep, and the company we should entertain when he got his advancement. My pride rose against this barefaced way of pointing out the contrast my married life was to present to my then dependent and inferior position. I suppressed my indignation; but, I showed her that her intention was not lost upon me, and I repaid her annoyances by affecting humility. What she described, would surely be a great deal too much honour for me, I would tell her. I was afraid I might not be able to support so great a change. Think of a mere governess, her daughter's governess, coming to that high distinction! It made her uneasy, and made them all uneasy, when I answered in this way. They knew that I fully understood her.

It was at the time when my troubles were at their highest, and when I was most incensed against my lover for his ingratitude in caring as little as he did for the innumerable distresses and mortifications I underwent for his account, that your dear friend, Mr. Gowan, appeared at the house. He had been intimate there for a long time, but had been abroad. He understood the state of things at a glance, and he understood me.

He was the first person I had ever seen in my life who had understood me. He was not in the house three times before I knew that he accompanied every movement of my mind. In his cold easy way with all of them, and with me, and with the whole subject, I saw it clearly. In his light protestations of admiration of my future husband, in his enthusiasm regarding our engagement and our prospects, in his hopeful congratulations on our future wealth and his despondent references to his own poverty—all equally hollow, and jesting, and full of mockery—I saw it clearly. He made me feel more and more resentful, and more and more contemptible, by always presenting to me everything that surrounded me, with some new hateful light upon it, while he pretended to exhibit it in its best aspect for my admiration and his own. He was like the dressed-up Death in the Dutch series; whatever figure he took upon his arm, whether it was youth or age, beauty or ugliness, whether he danced with it, sang it, played with it or prayed with it, he made it ghastly.

You will understand, then, that when your dear friend complimented me, he really condoled with me; that when he soothed me under my vexations, he laid bare every smarting wound I had; that when he declared my "faithful swain" to be "the most loving young fellow in the world, with the tenderest heart that ever beat," he touched my old misgiving that I was made ridiculous. These were not great services, you may say. They were acceptable to me, because they echoed my own mind, and confirmed my own knowledge. I soon began to like the society of your dear friend better than any other.

When I perceived (which I did, almost as soon) that jealousy was growing out of this, I liked this society still better. Had I not been subjected to jealousy, and were the endurances to be all mine? No. Let him know what it was! I was delighted that he should know it; I was delighted that he should feel keenly, and I hoped he did. More than that. He was tame in comparison with Mr. Gowan, who knew how to address me on equal terms, and how to anatomise the wretched people around us.

This went on, until the aunt, my Mistress, took it upon herself to speak to me. It was scarcely worth alluding to; she knew I meant

nothing; but, she suggested from herself, knowing it was only necessary to suggest, that it might be better if I were a little less companion-
able with Mr. Gowan.

I asked her how she could answer for what I meant? She could
always answer, she replied, for my meaning nothing wrong. I thanked
her, but I said I would prefer to answer for myself and to myself. Her
other servants would probably be grateful for good characters, but I
wanted none.

Our conversation followed, and induced me to ask her how she knew
that it was only necessary for her to make a suggestion to me, to have
it obeyed? Did she presume on my birth, or on my hire? I was not
bought, body and soul. She seemed to think that her distinguished
nephew had gone into a slave-market and purchased a wife.

It would probably have come, sooner or later, to the end to which
it did come, but she brought it to an issue at once. She told me, with
assumed commiseration, that I had an unhappy temper. On this repetition of the old wicked injury, I withheld no longer, but exposed to her
all I had known of her and seen in her, and all I had undergone
within myself since I had occupied the despicable position of being
engaged to her nephew. I told her that Mr. Gowan was the only relief
I had had in my degradation; that I had borne it too long, and that I
shook it off too late; but, that I would see none of them more. And I
never did.

Your dear friend followed me to my retreat, and was very droll on
the severance of the connexion; though he was sorry, too, for the excellent people (in their way the best he had ever met), and deplored
the necessity of breaking mere house-flies on the wheel. He protested
before long, and far more truly than I then supposed, that he was not
worth acceptance by a woman of such endowments, and such power
of character; but—well, well!—

Your dear friend amused me and amused himself as long as it suited
his inclinations; and then reminded me that we were both people of
the world; that we both understood mankind, that we both knew there
was no such thing as romance, that we were both prepared for going
different ways to seek our fortunes like people of sense, and that we
both foresaw that whenever we encountered one another again we
should meet as the best friends on earth. So he said, and I did not contradict him.

It was not very long before I found that he was courting his present wife, and that she had been taken away to be out of his reach. I hated her then, quite as much as I hate her now; and naturally, therefore, could desire nothing better than that she should marry him. But, I was restlessly curious to look at her—so curious that I felt it to be one of the few sources of entertainment left to me. I travelled a little: travelled until I found myself in her society, and in yours. Your dear friend, I think, was not known to you then, and had not given you any of those signal marks of his friendship which he has bestowed upon you.

In that company I found a girl, in various circumstances of whose position there was a singular likeness to my own, and in whose character I was interested and pleased to see much of the rising against swollen patronage and selfishness, calling themselves kindness, protection, benevolence, and other fine names, which I have described as inherent in my nature. I often heard it said, too, that she had "an unhappy temper." Well understanding what was meant by the convenient phrase, and wanting a companion with a knowledge of what I knew, I thought I would try to release the girl from her bondage and sense of injustice. I have no occasion to relate that I succeeded.

We have been together ever since, sharing my small means.

DJUNA BARNES
(1 8 9 3 –   )

Nightwood: Night Watch

The strangest "salon" in America was Nora's. Her house was couched in the center of a mass of tangled grass and weeds. Before it fell into Nora's hands the property had been in the same family two hundred years. It had its own burial ground, and a decaying chapel in which stood in tens and tens moldering psalm books, laid down some fifty years gone in a flurry of forgiveness and absolution.

It was the "paupers'" salon for poets, radicals, beggars, artists, and people in love; for Catholics, Protestants, Brahmins, dabblers in black magic and medicine; all these could be seen sitting about her oak table before the huge fire, Nora listening, her hand on her hound, the firelight throwing her shadow and his high against the wall. Of all that ranting, roaring crew, she alone stood out. The equilibrium of her nature, savage and refined, gave her bridled skull a look of compassion. She was broad and tall, and though her skin was the skin of a child, there could be seen coming, early in her life, the design that was to be the weatherbeaten grain of her face, that wood in the work; the tree coming forward in her, an undocumented record of time.

She was known instantly as a Westerner. Looking at her, foreigners remembered stories they had heard of covered wagons; animals going down to drink; children's heads, just as far as the eyes, looking in fright out of small windows, where in the dark another race crouched in ambush; with heavy hems the women becoming large, flattening the fields where they walked: God so ponderous in their minds that they could stamp out the world with him in seven days.

At these incredible meetings one felt that early American history was being reenacted. The Drummer Boy, Fort Sumter, Lincoln, Booth, all somehow came to mind; Whigs and Tories were in the air; bunting and its stripes and stars, the swarm increasing slowly and accurately on the hive of blue; Boston tea tragedies, carbines, and the sound of a boy's wild calling; Puritan feet, long upright in the grave, striking the earth again, walking up and out of their custom; the calk of prayers thrust in the heart. And in the midst of this, Nora.

By temperament Nora was an early Christian; she believed the word. There is a gap in "world pain" through which the singular falls continually and forever; a body falling in observable space, deprived of the privacy of disappearance; as if privacy, moving relentlessly away, by the very sustaining power of its withdrawal kept the body eternally moving downward, but in one place, and perpetually before the eye. Such a singular was Nora. There was some derangement in her equilibrium that kept her immune from her own descent.

Nora had the face of all people who love the people—a face that would be evil when she found out that to love without criticism is to be betrayed. Nora robbed herself for everyone; incapable of giving herself warning, she was continually turning about to find herself diminished. Wandering people the world over found her profitable in that she could be sold for a price forever, for she carried her betrayal money in her own pocket.

Those who love everything are despised by everything, as those who love a city, in its profoundest sense, become the shame of that city, the *détraqués*, the paupers; their good is incommunicable, outwitted, being the rudiment of a life that has developed, as in man's body are found evidences of lost needs. This condition had struck even into Nora's house; it spoke in her guests, in her ruined gardens where she had been wax in every work of nature.

Whenever she was met, at the opera, at a play, sitting alone and apart, the program face down on her knee, one would discover in her eyes, large, protruding and clear, that mirrorless look of polished metals which report not so much the object as the movement of the object. As the surface of a gun's barrel, reflecting a scene, will add to the image the portent of its construction, so her eyes contracted and fortified the play before her in her own unconscious terms. One sensed in the way she held her head that her ears were recording Wagner or Scarlatti, Chopin, Palestrina, or the lighter songs of the Viennese school, in a smaller but more intense orchestration.

And she was the only woman of the last century who could go up a hill with the Seventh Day Adventists and confound the seventh day—with a muscle in her heart so passionate that she made the seventh day immediate. Her fellow worshipers believed in that day and the end of the world out of a bewildered entanglement with the six days preceding it; Nora believed for the beauty of that day alone. She was by fate one of those people who are born unprovided for, except in the provision of herself.

One missed in her a sense of humor. Her smile was quick and definite, but disengaged. She chuckled now and again at a joke, but it was the amused grim chuckle of a person who looks up to discover that they have coincided with the needs of nature in a bird.

Cynicism, laughter, the second husk into which the shucked man crawls, she seemed to know little or nothing about. She was one of those deviations by which man thinks to reconstruct himself.

To "confess" to her was an act even more secret than the communication provided by a priest. There was no ignominy in her; she recorded without reproach or accusation, being shorn of self-reproach or self-accusation. This drew people to her and frightened them; they could neither insult nor hold anything against her, though it embittered them to have to take back injustice that in her found no foothold. In court she would have been impossible; no one would have been hanged, reproached or forgiven, because no one would have been "accused." The world and its history were to Nora like a ship in a bottle; she herself was outside and unidentified, endlessly embroiled in a preoccupation without a problem.

Then she met Robin. The Denckman circus, which she kept in touch with even when she was not working with it (some of its people were visitors to her house), came into New York in the fall of 1923. Nora went alone. She came into the circle of the ring, taking her place in the front row.

Clowns in red, white and yellow, with the traditional smears on their faces, were rolling over the sawdust, as if they were in the belly of a great mother where there was yet room to play. A black horse, standing on trembling hind legs that shook in apprehension of the raised front hoofs, his beautiful ribboned head pointed down and toward the trainer's whip, pranced slowly, the foreshanks flickering to the whip. Tiny dogs ran about trying to look like horses, then in came the elephants.

A girl sitting beside Nora took out a cigarette and lit it; her hands shook and Nora turned to look at her; she looked at her suddenly because the animals, going around and around the ring, all but climbed over at that point. They did not seem to see the girl, but as their dusty eyes moved past, the orbit of their light seemed to turn on her. At that moment Nora turned.

The great cage for the lions had been set up, and the lions were walking up and out of their small strong boxes into the arena. Ponderous and furred they came, their tails laid down across the floor, dragging and heavy, making the air seem full of withheld strength. Then as one powerful lioness came to the turn of the bars, exactly opposite the girl, she turned her furious great head with its yellow eyes afire and went down, her paws thrust through the bars and, as she regarded the girl, as if a river were falling behind impassable heat, her eyes flowed in tears that never reached the surface. At that the girl rose straight up. Nora took her hand. "Let's get out of here!" the girl said, and still holding her hand Nora took her out.

In the lobby Nora said, "My name is Nora Flood," and she waited.

After a pause the girl said, "I'm Robin Vote." She looked about her distractedly. "I don't want to be here." But it was all she said; she did not explain where she wished to be.

She stayed with Nora until the midwinter. Two spirits were working in her, love and anonymity. Yet they were so "haunted" of each other that separation was impossible.

Nora closed her house. They traveled from Munich, Vienna and Budapest into Paris. Robin told only a little of her life, but she kept repeating in one way or another her wish for a home, as if she were afraid she would be lost again, as if she were aware, without conscious knowledge, that she belonged to Nora, and that if Nora did not make it permanent by her own strength, she would forget.

Nora bought an apartment in the Rue du Cherche-Midi. Robin had chosen it. Looking from the long windows one saw a fountain figure, a tall granite woman bending forward with lifted head; one hand was held over the pelvic round as if to warn a child who goes incautiously.

In the passage of their lives together every object in the garden, every item in the house, every word they spoke, attested to their mutual love, the combining of their humors. There were circus chairs, wooden horses bought from a ring of an old merry-go-round, Venetian chandeliers from the Flea Fair, stage drops from Munich, cherubim from Vienna, ecclesiastical hangings from Rome, a spinet from England, and a miscellaneous collection of music boxes from many countries; such was the museum of their encounter, as Felix's hearsay house had been testimony of the age when his father had lived with his mother.

When the time came that Nora was alone most of the night and part of the day, she suffered from the personality of the house, the punishment of those who collect their lives together. Unconsciously at first, she went about disturbing nothing; then she became aware that her soft and careful movements were the outcome of an unreasoning fear—if she disarranged anything Robin might become confused—might lose the scent of home.

Love becomes the deposit of the heart, analogous in all degrees to the "findings" in a tomb. As in one will be charted the taken place of the body, the raiment, the utensils necessary to its other life, so in the heart of the lover will be traced, as an indelible shadow, that which he loves. In Nora's heart lay the fossil of Robin, intaglio of her identity, and about it for its maintenance ran Nora's blood. Thus the body of Robin could never be unloved, corrupt or put away. Robin was now beyond timely changes, except in the blood that animated her. That she could be spilled of this fixed the walking image of Robin in appalling apprehension on Nora's mind—Robin alone, crossing streets, in danger. Her mind became so transfixed that, by the agency of her fear, Robin seemed enormous and polarized, all catastrophes ran toward her, the magnetized predicament; and crying out, Nora would wake from sleep, going back through the tide of dreams into which her anxiety had thrown her, taking the body of Robin down with her into it, as the ground things take the corpse, with minute persistence, down into the earth, leaving a pattern of it on the grass, as if they stitched as they descended.

Yes now, when they were alone and happy, apart from the world in their appreciation of the world, there entered with Robin a company unaware. Sometimes it rang clear in the songs she sang, sometimes Italian, sometimes French or German, songs of the people, debased and haunting, songs that Nora had never heard before, or that she had never heard in company with Robin. When the cadence changed, when it was repeated on a lower key, she knew that Robin was singing of a life that she herself had no part in; snatches of harmony as telltale as the possessions of a traveler from a foreign land; songs like a practiced whore who turns away from no one but the one who loves her. Sometimes Nora would sing them after Robin, with the trepidation of a foreigner repeating words in an unknown tongue, uncertain of what they may mean. Sometimes unable to endure the melody that told so much and so little, she would interrupt Robin with a question. Yet more distressing would be the moment when, after a pause, the song would be taken up again from an inner room where Robin, unseen, gave back an echo of her unknown life more nearly tuned to its origin. Often the song would stop altogether, until unthinking, just as she was leaving the house, Robin would break out again in anticipation, changing the sound from a reminiscence to an expectation.

Yet sometimes, going about the house, in passing each other, they would fall into an agonized embrace, looking into each other's face, their two heads in their four hands, so strained together that the space that divided them seemed to be thrusting them apart. Sometimes in these moments of insurmountable grief Robin would make some movement, use a peculiar turn of phrase not habitual to her, innocent of the betrayal, by which Nora was informed that Robin had come from a world to which she would return. To keep her (in Robin there was this tragic longing to be kept, knowing herself astray) Nora knew now that there was no way but death. In death Robin would belong to her. Death went with them, together and alone; and with the torment and catastrophe, thoughts of resurrection, the second duel.

Looking out into the fading sun of the winter sky, against which a little tower rose just outside the bedroom window, Nora would tabulate by the sounds of Robin dressing the exact progress of her toilet; chimes of cosmetic bottles and cream jars; the faint perfume of hair heated under the electric curlers; seeing in her mind the changing direction taken by the curls that hung on Robin's forehead, turning back from the low crown to fall in upward curves to the nape of the neck, the flat uncurved back head that spoke of some awful silence. Half narcoticized by the sounds and the knowledge that this was in preparation for departure, Nora spoke to herself: "In the resurrection, when we come up looking backward at each other, I shall know you only of all that company. My ear shall turn in the socket of my head; my eyeballs loosened where I am the whirlwind about that cashed expense, my foot stubborn on the cast of your grave." In the doorway Robin stood. "Don't wait for me," she said.

In the years that they lived together, the departures of Robin became slowly increasing rhythm. At first Nora went with Robin; but as time passed, realizing that a growing tension was in Robin, unable to endure the knowledge that she was in the way or forgotten, seeing Robin go from table to table, from drink to drink, from person to person, realizing that if she herself were not there Robin might return to her as the one who, out of all the turbulent night, had not been lived through, Nora stayed at home, lying awake or sleeping. Robin's absence, as the night drew on, became a physical removal, insupportable and irreparable. As an amputated hand cannot be disowned because it is experiencing a futurity, of which the victim is its forebear, so Robin was an amputation that Nora could not renounce. As the wrist longs, so her heart longed, and dressing she would go out into the night that she might be "beside herself," skirting the café in which she could catch a glimpse of Robin.

Once out in the open Robin walked in a formless meditation, her hands thrust into the sleeves of her coat, directing her steps toward that night life that was a known measure between Nora and the cafés. Her meditations, during this walk, were a part of the pleasure she expected to find when the walk came to an end. It was this exact distance that kept the two ends of her life—Nora and the cafés—from forming a monster with two heads.

Her thoughts were in themselves a form of locomotion. She walked with raised head, seeming to look at every passerby, yet her gaze was anchored in anticipation and regret. A look of anger, intense and hurried, shadowed her face and drew her mouth down as she neared her company; yet as her eyes moved over the façades of the buildings, searching for the sculptured head that both she and Nora loved (a Greek head with shocked protruding eyeballs, for which the tragic mouth seemed to pour forth tears), a quiet joy radiated from her own eyes; for this head was remembrance of Nora and her love, making the anticipation of the people she was to meet set and melancholy. So, without knowing she would do so, she took the turn that brought her into this particular street. If she was diverted, as was sometimes the case, by the interposition of a company of soldiers, a wedding or a funeral, then by her agitation she seemed a part of the function to the persons she stumbled against, as a moth by his very entanglement with the heat that shall be his extinction is associated with flame as a component part of its function. It was this characteristic that saved her from being asked too sharply "where" she was going; pedestrians who had it on the point of their tongues, seeing her rapt and confused, turned instead to look at each other.

The doctor, seeing Nora out walking alone, said to himself, as the tall black-caped figure passed ahead of him under the lamps, "There goes the dismantled—Love has fallen off her wall. A religious woman," he thought to himself, "without the joy and safety of the Catholic faith, which at a pinch covers up the spots on the wall when the family portraits take a slide; take that safety from a woman," he said to himself, quickening his step to follow her, "and love gets loose and into the rafters. She sees her everywhere," he added, glancing at Nora as she passed into the dark. "Out looking for what she's afraid to find— Robin. There goes mother of mischief, running about, trying to get the world home."

Looking at every couple as they passed, into every carriage and car, up to the lighted windows of the houses, trying to discover not Robin any longer, but traces of Robin, influences in her life (and those which were yet to be betrayed), Nora watched every moving figure for some gesture that might turn up in the movements made by Robin; avoiding the quarter where she knew her to be, where by her own movements the waiters, the people on the terraces, might know that she had a part in Robin's life.

Returning home, the interminable night would begin. Listening to the faint sounds from the street, every murmur from the garden, an unevolved and tiny hum that spoke of the progressive growth of noise that would be Robin coming home, Nora lay and beat her pillow without force, unable to cry, her legs drawn up. At times she would get up and walk, to make something in her life outside more quickly over, to bring Robin back by the very velocity of the beating of her heart. And walking in vain, suddenly she would sit down on one of the circus chairs that stood by the long window overlooking the garden, bend forward, putting her hands between her legs, and begin to cry, "Oh, God! Oh, God! Oh, God!" repeated so often that it had the effect of all words spoken in vain. She nodded and awoke again and began to cry before she opened her eyes, and went back to the bed and fell into a dream which she recognized; though in the finality of this version she knew that the dream had not been "well dreamt" before. Where the dream had been incalculable, it was now completed with the entry of Robin.

Nora dreamed that she was standing at the top of a house, that is, the last floor but one—this was her grandmother's room—an expansive, decaying splendor; yet somehow, though set with all the belongings of her grandmother, was as bereft as the nest of a bird which will not return. Portraits of her great-uncle, Llewellyn, who died in the Civil War, faded pale carpets, curtains that resembled columns from their time in stillness—a plume and an inkwell—the ink faded into the quill; standing, Nora looked down into the body of the house, as if from a scaffold, where now Robin had entered the dream, lying among a company below. Nora said to herself, "The dream will not be dreamed again." A disc of light, which seemed to come from someone or thing standing behind her and which was yet a shadow, shed a faintly luminous glow upon the upturned still face of Robin, who had the smile of an "only survivor," a smile which fear had married to the bone.

From round about her in anguish Nora heard her own voice saying, "Come up, this is Grandmother's room," yet knowing it was impossible because the room was taboo. The louder she cried out the farther away went the floor below, as if Robin and she, in their extremity, were a pair of opera glasses turned to the wrong end, diminishing in their painful love; a speed that ran away with the two ends of the building, stretching her apart.

This dream that now had all its parts had still the former quality of never really having been her grandmother's room. She herself did not seem to be there in person, nor able to give an invitation. She had wanted to put her hands on something in this room to prove it; the dream had never permitted her to do so. This chamber that had never been her grandmother's, which was, on the contrary, the absolute opposite of any known room her grandmother had ever moved or lived in, was nevertheless saturated with the lost presence of her grandmother, who seemed in the continual process of leaving it. The architecture of dream had rebuilt her everlasting and continuous, flowing away in a long gown of soft folds and chin laces, the pinched gatherings that composed the train taking an upward line over the back and hips in a curve that not only bent age but fear of bent age demands.

With this figure of her grandmother who was not entirely her recalled grandmother, went one of her childhood, when she had run into her at the corner of the house—the grandmother who, for some unknown reason, was dressed as a man, wearing a billycock and a corked moustache, ridiculous and plump in tight trousers and a red waistcoat, her arms spread, saying with a leer of love, "My little sweetheart!"— her grandmother "drawn upon" as a prehistoric ruin is drawn upon, symbolizing her life out of her life, and which now appeared to Nora as something being done to Robin, Robin disfigured and eternalized by the hieroglyphics of sleep and pain.

Waking, she began to walk again, and looking out into the garden in the faint light of dawn, she saw a double shadow falling from the statue, as if it were multiplying, and thinking perhaps this was Robin, she called and was not answered. Standing motionless, straining her eyes, she saw emerge from the darkness the light of Robin's eyes, the fear in them developing their luminosity until, by the intensity of their double regard, Robin's eyes and hers met. So they gazed at each other. As if that light had power to bring what was dreaded into the zone of their catastrophe, Nora saw the body of another woman swim up into the statue's obscurity, with head hung down, that the added eyes might not augment the illumination; her arms about Robin's neck, her body pressed to Robin's, her legs slackened in the hang of the embrace.

Unable to turn her eyes away, incapable of speech, experiencing a sensation of evil, complete and dismembering, Nora fell to her knees, so that her eyes were not withdrawn by her volition, but dropped from their orbit by the falling of her body. Her chin on the sill, she knelt, thinking, "Now they will not hold together," feeling that if she turned away from what Robin was doing, the design would break and melt back into Robin alone. She closed her eyes, and at that moment she knew an awful happiness. Robin, like something dormant, was protected, moved out of death's way by the successive arms of women; but as she closed her eyes, Nora said "Ah!" with the intolerable automatism of the last "Ah!" in a body struck at the moment of its final breath.

Only one story, because Patrick White's "On the Balcony" does it so well—the slow, spreading sadness of shriveling desire, the retreat from hope to fantasy that was the common lot of the woman who did not marry in a traditional society where no other form of adult femininity (except the convent in Catholic countries) was recognized.

PATRICK WHITE
(1912–   )

On the Balcony

"He will be here any moment now," Alexandra said. "He is very punctual."

"Yes, yes," said Fanny. "Of course."

Admitting her own weaknesses, she was now all open to reproach. She knew that she liked to lie too long in the bath, hypnotized by the sight of herself under water, or afterward to sit on the edge and sing, while the little drops of water trembled and glistened on her white skin, the moment before extinction. But now she was all self-reproach, and the bedroom full of hairpins and open drawers and the subtle sound of silk.

"Yes," she admitted. "But you accuse me, Alexandra, always," she said.

"It is habit," yawned Alexandra.

She sat at the bedroom window darning a stocking, and her movements broke the indolence of Sunday afternoon. It made her feel slightly superior. Her expression was both domesticated and mild. But Alexandra Papaioannou was not perpetually mild. Her face had bones. From time to time she looked away from her thread down out of the window, ready to pounce on the figure of Mr. Angelopoulos on his first appearance in the street. And Fanny dreaded this, the unadorned announcement by her sister, like the messenger in a tragedy.

Mrs. Papaioannou sometimes said that if Fanny was soft, Alexandra was hard. But on the whole they were closely united, the Papaioannous, from social and economic necessity as much as anything else. Their existence was the laborious one of keeping up appearances. Living in a small apartment, though at an excellent address (from its position on the hill the house looked down on the rest of Athens), Mrs. Papaioannou yearned after past glories. Unconsciously her life was divided into Before and After the Catastrophe. Her world, the real tangible world, had ceased with Smyrna, and now the best she could do was chafe back a melancholy reflection of reality onto the face of life. The world of Mrs. Papaioannou was a world of gold and Anatolian estates, where the evening filled with sea breezes and the scent of purple stocks, and the sun set across the gulf. She was made for ease, and morning gowns, and *laisser-aller,* whereas Athens sometimes frightened her. Its air quivered with a dry and brilliant malice. Like a big downy Smyrna peach, she had fallen on the stones of Attica and bruised.

But her girls were more adaptable. Their ship still sailed with the wind of possibility. Though make haste, make haste, their mother would often have liked to say, the rocks begin to appear in the middle thirties. For Fanny was thirty-five, and Alexandra just far enough behind to preserve the decencies.

But the wind still stood fair, and on a Sunday afternoon in spring the ship sailed, at any rate for Fanny. She could feel it in the air, on her face, trying a necklace round her throat, and deciding that perhaps simplicity, yes, simplicity after all had a quiet strength of its own.

"He is coming," said Alexandra out of the window. "With a bunch of anemones. He has taken off his hat. He is hot."

"Naturally," Fanny said. "It's the hill."

"And, oh, Fanny, he is getting very thin on top."

There were some things Fanny refused to hear. It was inevitable, sharing a room, living on top of each other in a small flat. In the circumstances she concentrated on the anemones. She would say something pretty, and simple, and sincere, holding them to her breast. Because she could still get away with this sort of thing, like an ingenue in muslin in a French film. She narrowed her eyes a little at her own reflection in the glass, all red and white and black. There was nothing hard about Fanny Papaioannou, her mother was right. She melted like a summer evening by the sea.

"He is here," said Kyriaki.

"Who?" said Fanny Papaioannou. "The butcher? The baker? Who?"

She spoke sharply. She would take out on Kyriaki the annoyance she felt with Alexandra. And Kyriaki squinted because she did not understand, she could never understand the language of young ladies who read novels in French.

"Why, the Teacher, of course," she said.

Resentment made her sweat and finger her dress.

"How many times have I told you," said Fanny, "never, *never* to refer to Mr. Angelopoulos as the Teacher?"

Sometimes her lips grow thin, thought Alexandra, darning a stocking.

"But he teaches," said Kyriaki.

"Yes," said Fanny Papaioannou. "But—"

She refused to spoil her composure in battle, particularly as she knew she would lose. For Kyriaki had the strength as well as the limitations of extreme simplicity. Already she was daring to smile. She stood in the doorway smiling, a brown girl from a village, showing a gold tooth.

"Well," she said. "Anyway the gentleman is here, Miss Fanny; I put him on the balcony."

And then she went. And it was time. It was time to go out and play the word-perfect part that frightened you nevertheless, however often rehearsed, that first entrance under the full glare of a situation. Fanny Papaioannou felt suddenly tender.

"Pray for me, dear Alexandra," she said, as she kissed her sister on the cheek.

Mr. Angelopoulos sat on the balcony fanning himself with *Elevthero Vima*. He was a gray man with pale eyes, of indeterminate age, in fact everything about him was indeterminate, though excellent, everything in the best of taste. Highly thought of, too, at the Byzantine School where he taught. The headmaster would ask his advice in solving problems of policy, when even a silence from Angelopoulos had the sustaining quality of moral support.

Now he got out of his chair, unfolding his rather thin legs, and offering a bunch of anemones said:

"A touch of spring, Miss Fanny."

"They are exquisite. Quite exquisite," she said.

She heard herself through her own caught breath, and knew with a twinge of shame that this was a word she never used, the kind of word that Elias Angelopoulos invariably forced into her mouth because she so much wanted to share his own excellent taste. As he looked at her out of his pale eyes, as she settled her dress humbly on an upright chair, she knew she was unworthy.

Mr. Angelopoulos had begun to fan himself again with the copy of *Elevthero Vima*.

"The hill is very steep," she said quickly. "If one isn't accustomed to it, that is, three or four times a day—then, of course, it becomes part of one's daily life like, like anything else. Mama says we have grown into mountaineers."

She heard her laugh. She did not like her own laugh. It came out too girlishly sometimes, in a high *chi-chi-chi*.

"Every hill depends on its summit," said Angelopoulos, because he always said the right thing.

She looked at him with an expression of admiration that was almost hate, though why she would not have known. She stroked the heads of the anemones with the desperation of escaping time. Because it did escape, slippery as eels. Sedate enough in the ordinary run of events, there were still the moments of illumination when you realized. The years and the Papaioannou girls. The horrifying thought. Eleven months in the year the Papaioannou girls indulged in the pursuit of husbands in Athens. One month of summer they spent with Aunt Katie Dimitriou on Aiyina, and relaxed. Aunt Katie was a spinster who offered advice. Sitting on the balcony with Elias Angelopoulos, Fanny longed for even the unprofitable advice of Aunt Katie Dimitriou. Aunt Katie had once had a proposal from a young man, an excellent young man, but a young man of a lower class. "We of the aristocracy . . ." Aunt Katie liked to say to her sister. So the young man went away and married a girl he first saw in a photograph, and became a merchant with four children in Piraeus. "You can well imagine," Aunt Katie said, "dried codfish and olive oil." "Yes, Katie," Mrs. Papaioannou used to say, summer afternoons on Aiyina, "Yes Katie," she would say, "but we are not all gifted with ideals." Summer afternoons on Aiyina, after too much stuffed marrow or pilaff, or milk lamb, Mrs. Papaioannou preferred to be left to her own indolence, and to the control of what she had to admit was an attack of wind. But now Fanny would have given much to have had Aunt Katie on the balcony with an offering of strength of mind.

Instead the girl came, and this at least was a release. The presence of Kyriaki with the tray held stiffly in her brown hands that trembled, *elle a des tremblements de vierge,* Mrs. Papaioannou used to say, and the little coffee cups rattled on the tray. Kyriaki trembled and smiled. Somewhere inside her a gust of laughter waited to blow her to perdition.

"Please," said Fanny Papaioannou.

"Wild figs!" said Mr. Angelopoulos.

"Yes," she said hopelessly.

"I eat very little," he said. "You know that my stomach doesn't allow me. One must always consider one's health."

"Yes," she said. "Of course."

"But wild figs!"

She breathed. Her feelings would not allow her to interpret the expression on his face, poised above the figs, as one of mild greed. From certain angles his mustache was most distinguished, gray, and glistening, and brushed.

Mr. Angelopoulos spooned a fig out of the little glass dish, making it a long gesture of deep aesthetic approval.

"Edible jade," he murmured, and popped it under his mustache, all but the two or three drops of syrup that spilled upon his gray leg.

She began to talk very quickly. She would not notice.

"They have a quality that is different," she said, "a *country* flavor. After the wild ones, there is something rather insipid about the domesticated fig. But one feels, one feels that the little, wild, green figs have so much *character*," she said.

"Exactly," said Mr. Angelopoulos.

He had taken his handkerchief and begun to dab at his trouser leg.

Incidents like these gnawed long afterward at his mind. His mother, with whom he lived, would have taken some warm water and a sponge. But here he sat holding a coffee cup and dabbing with a handkerchief, and his day was spoiled. He heard with annoyance the rattle of the dish of figs as the brown girl took away the tray.

The life of Elias Angelopoulos had a pattern that he had composed at an early age. It was as neat and formal as the white façade of the Byzantine School at which he taught, as orderly as the hours in which he gave lessons in French language and literature to older children. The Byzantine School was in every way circumspect. It was not the policy of the school to encourage the children to take political sides and clamor for revenge, but as the children were Greeks, in time they usually did. Then it pained Elias Angelopoulos, in the school that he loved, to hear the scuffle of feet on the gravel and hot words falling among the laurel bushes in the break. Then he pretended not to hear, it was his habit in the face of the more disorderly aspects of life; he would concentrate on the laurel bushes to which he was attached, because the laurel, he felt, was a neat and unemotional shrub. The enthusiasms of Mr. Angelopoulos were largely literary. He read the French classics you would have said with passion for something done so often, but this was a word you could not use in connection with Angelopoulos. Rather he read the French classics applauding his own good taste. And out of this good taste formed, not so much a love, as an affection for Vigny, and Sully Prudhomme and Corneille. For a time, in his youth, he admired Racine, but only for a time, there was an uncomfortable lack of compromise in a relationship with Racine, like looking at the frosted glass of a bathroom window and suddenly finding it was not frosted at all. This was what Elias Angelopoulos had discovered; he remembered one day he was reading *Phèdre,* and suddenly it was as clear as glass, peering out from among the stiffness and the flowers, a face softening into flesh and a voice stripped naked by the words of doom! *De l'amour j'ai toutes les fureurs.* For many years Angelopoulos had contemplated this revelation and rejected with a warm distaste everything that it might imply.

"Sometimes I feel I would like to be a gypsy," Fanny Papaionnou was saying.

He coughed and put down his coffee cup. He smiled indulgently at the fancies of young girls. But she was pretty, he admitted, experiencing a slight sensation in his gray thighs. She had sat forward in her chair, an elbow on the balcony ledge, a hand supporting her cheek, in a way that she knew suited her; by day the sun would accentuate the blueness of her hair and by night the moon the whiteness of her skin. Now it was dusk, and the mauve shadows of the hills were on her cheek.

"Do you believe in fortunes, Elia?" she asked.

Kyriaki stood just inside the salon. She snickered and waited and touched the curtains.

Mr. Angelopoulos cleared his throat with the air of a person who has made up his mind he is not going to be caught.

"The irrational," he said. This was going to be a remark of some importance, "the irrational adds a certain color to existence, if nothing else."

"Oh, I love them," she said. "Fortunes, I mean. Just as a joke. It's amusing to see."

She clasped her hands very tightly, as if she were afraid they were going to fly away, two white birds, into the falling light.

"Just to see," she said. "For fun. Kyriaki," she called. "Kyriaki has magic. Kyriaki looks into the cups."

Mr. Angelopoulos laughed, indulgently, for the fancies of young girls.

Kyriaki giggled. She smoothed her skirt with trembling hands, because now her moment had come, the moment of the cups.

"Now, Kyriaki," said Fanny. "You shall tell us many important things."

"If it's there, it's there," said Kyriaki.

"You see, Elia, she believes!"

"Of course she believes," said Angelopoulos.

In his heart he felt he ought to protest.

"Now, Miss Fanny, you make fun of me!"

The girl giggled in time with the rattle of the cup that she picked up, held elegantly, in the way she had seen ladies do, with a brown and stumpy hand.

"Ah, here's many things," she said. "Oh, sir, sir!"

She held her head on one side, the brown hair falling. She had the rapt expression of a half-ashamed conviction. Sometimes Kyriaki is almost simple, Fanny felt.

"Many things," said Kyriaki. "Look, sir, Mr. Elia, the open road. That is a journey," she said. "And look, oh tears, bitter things. And money. Dear, dear!"

Fanny Papaioannou listened to another fortune by Kyriaki. And the tears, and the journeys, and the journeys, and the tears. She knew it almost off by heart. But she waited and bit her lip.

"Ah, now, sir, here! Who would have thought! A dark lady. Oh, sir!"

"A dark lady, however," said Mr. Angelopoulos. "Ha, ha, ha!" he laughed. "A dark lady, Kyriaki!"

And Fanny Papaioannou laughed. Her laughter came out in that key, only under the stress of emotion, which she always immediately regretted, hearing her own high breathless *chi-chi-chi*.

"She makes too much of this *chi-chi-chi*," said Alexandra in the bedroom where, under a hard light, she sat darning a stocking. Younger than Fanny, she was also more practical and collected. She sat darning her stocking and might have been weaving a destiny.

"And tears for the lady," said Kyriaki. "See? A bend in the road. That means love never runs straight."

She turned the cup solemnly, a dark world revolving, from which all things flowed, whether the private destiny or the warm darkness of the night. Because the light had died now. All along the Saronic Gulf it was as pale as doves. The street had a liveliness of its own, full of the barking of dogs, and the inconsequential singing, and the sharp electric squares that each framed a bosom and a pair of arms. But all dependent, waiting on the mystic revolution of a cup.

Oh, but this was too much, felt Fanny. She was waiting. She would hear.

"Come, what else, Kyriaki?"

Kyriaki giggled. It began to seize her. She was possessed by some private vision. She was shaken by a great laughter that came up out of her stomach and almost made her topple with the cup.

"Well?" said Angelopoulos, as he sat forward, forgetting his shame.

"No, no," screamed Kyriaki.

Now she was almost devilish, and several of her teeth were gold.

"You are a stupid girl," Fanny said. "Quite an idiot. Tell us, Kyriaki. I insist."

"Oh, Miss Fanny," said Kyriaki, and her voice got low, sucked back into her stomach on a wave of laughter. "Oh, Miss Fanny, you will never forgive. But it's there. It's there. A bed!"

That extinguished her, she was done, she had to retreat, and her laughter rolled back from the kitchen after she had gone, together with the shuffling of the cups.

Angelopoulos laughed a blameless, a colorless laugh.

"Really," said Fanny. "These peasants, these Greeks. Give them an inch," she said.

Now the thread was broken. She was moist with shame, glad that she could not see his eyes, the pale distinguished eyes that sometimes gave her a feeling as if, as if she had opened a volume of belles-lettres that she respected but would never read. She drew her breath. She would spin. Not all the forces of evil would destroy her. Under her flesh the bones were hard. She tautened the muscles of her neck.

But the night was against her, the night that was full of singing, and people laughing in the street. Under the ragged acacia tree on the corner sat the girl and the two soldiers. The light from the doorway fell on her frizzed head, as they sat drinking, the three, they clinked glasses, the frizzed girl with the wet mouth and the two brown-faced mustachy soldiers, eating little pieces of white cheese.

To raise it onto a higher plane, determined Fanny Papaioannou.

"Sometimes I feel so very much out of touch with things," she said aloud. "I mean, the essential things. It is so very important, don't you think, to identify oneself with the simple things? Sometimes I look at a piece of bread and see it for the first time. And I could cry. Because there seems to be nothing more beautiful than bread."

Mr. Angelopoulos cleared his throat.

"You have something of the creative spirit, Miss Fanny," he said.

"I?" and she laughed bitterly. "I haven't the humility," she said.

She liked that. She stroked her own arms in the dark, and her flesh was soft and satisfied.

Down under the acacia tree the frizzed girl clinked glasses with the two soldiers, of whom the smaller, very squat and black, his shirt open on his black chest, now leaned across and bit the lobe of her ear, and the frizzed girl shook him off her cheek, and began to sing a song about old photographs, their passion and their pain.

"If we can humble ourselves. Ah, yes, that is the question," said Mr. Angelopoulos.

He sidled uneasily, about to give birth to a confidence.

"I confess I dabble a little myself, Miss Fanny. Call it presumption if you like. A few sketches," he said, "a few sketches for a novel."

"Tell me," she said earnestly.

She sat forward and rested her face on her hands. She could not give him too much. She had almost dared to touch his arm.

"Yes," said Mr. Angelopoulos. "A novel on a social theme."

"Ah," she said, knowingly, and with such an inflection of sympathy she might have heard of an accident, or people slaughtered by Bulgarians.

Her song finished, the frizzed girl was shaking the crumbs from her skirt. She began to walk down the street, garlanded with soldiers' arms, and the darkness swallowed their three lives, making them part of its own.

"But it has endless possibilities!" said Fanny Papaioannou.

"If one has the strength," sighed Mr. Angelopoulos. "The humility and the strength."

He said it with the great simplicity of someone who has set himself apart.

"Oh, but you will, you will," said Fanny Papaioannou.

And she had the courage to touch his arm. She let her hand lie on his arm, and she closed her eyes, because it was an important moment. She regretted all the girlish things she had said that evening, before choosing the Higher Plane. She would give him all that she had inside her. Together they would create a great work, a work with a message to the Greek people. If he would let her. And he must. She dared to tighten the fingers on his arm.

Up the street blew a dark, a passionate gust.

Fanny Papaioannou opened her eyes and looked out, out of the private world of her own exaltation, saw in the room a street away, that somehow she had never noticed, the large woman in the pink kimono. The large woman stood in the electric light against the trellised paper. She faced into the room talking to the man on the brass bed, the man in braces, the man with the egg head. And all the time talking, talking, the large woman held the kimono wrapped to her body, as if trying to control its vast waves, to pour the sea into a pint pot. "Well, well?" said Fanny Papaioannou. There was no reason why the fat woman should not talk, but you felt, fascinated, watching the square of yellow light in which the figures stood out, that out of the great woman's mouth would come the ultimate answer. And Fanny was both fascinated and afraid.

"You have the gift of sympathy, Fanny," Mr. Angelopoulos was saying.

She felt the moment of crisis, the nearness and farness, because at the same time she had to admit that his voice had the calm agreeable tone of someone speaking an epitaph for somebody decently dead. And as for his arm, she might have been touching cloth.

"I hate myself," she said, and suddenly she meant it.

But above all she hated the large woman in the pink kimono, who was wheedling the man in braces, trying to entice him into her vast folds. The pink woman was as gigantic as the night. She bent and kissed the egg head. She stooped and took the egg's arms and incorporated them in her pink embrace. It was obvious, felt Fanny, there was no moral sanction in such an embrace at such an hour. And Angelopoulos, she could tell, she could sense it without a sideways glance, had become a surprised, an apprehensive stare.

Mr. Angelopoulos in fact removed his arm. He was talking about the time.

"I don't like to leave my mother. Her health is delicate," he said. "The bile."

"Must you?" she answered, only just.

Yes, the night had conspired. She could almost hear the shrieking of the brass bed. And the vast woman flowed. Her pink arms flew. And then the egg was laughing. You could see his stomach move. And the pink woman laughed, laughing into the room. The pink woman took, it was happening, there was no avoiding the moment, took off her pink gown, and stood as she was, and her behind shook with the immensity of her laughter.

The chair was scraping the balcony.

He would come on Sunday, he said.

"Yes," sighed Fanny Papaioannou.

But she avoided his discreet eyes. She was numbed by the closeness of reality. Go, she would like to have said, go. She heard the cracking of the pistol, the chalk-faced clown she had once seen, in green baize, who shot the clay pipe from out of his wife's mouth.

Anyway, she sighed, nursing her numbness now in the darkness, in the silence that he left behind.

In the kitchen Kyriaki felt melancholy, now that the excitement was over, now that it was dark. She drew her finger through the coffee dregs. She began to sing a sad song about a brigand who cut his mistress into several pieces and threw them over a precipice, somewhere in Epirus.

Fanny went into the bedroom.

"My hair is terrible, terrible," she said.

The rouge stood up on her cheeks in hard little patches.

"And what have you got to tell?" asked Alexandra, who sat at the window watching the last of Angelopoulos.

"Nothing today," Fanny said.

"But he will come on Sunday," said Alexandra.

She took the thread of darning silk. It had been going on so long now. Fanny and Angelopoulos. Every Sunday for three years. Except of course the month in summer they spent with Aunt Katie Dimitriou, on Aiyina. Alexandra Papaioannou took the silk thread, and with a certain air of finality, neatly knotted the end.

The Old Woman

Considering the number of women who died either in childbirth or prematurely from the weakening effects of repeated pregnancies, small wonder that those who survived to be old women were looked upon quite often as witches or as wise women, earth mothers possessed of remarkable powers. Aleksandr Solzhenitsyn's Matryona (in his story "Matryona's House") is a contemporary version of the wise old woman in touch with eternal verities, and Katherine Anne Porter's grandmother and her maid in "The Old Order" are two others. Then there are the images of stoic endurance, like Maria in Frank Waters' People of the Valley and Phoenix Jackson in Eudora Welty's story "A Worn Path," or the celebration of enduring joie de vivre in Vittorini's "La Garabaldina."

To the witch, earth mother, and wise old woman we have added the twinkly white-haired grandmother with spectacles and quavering voice who owes more to cake-mix commercials and television comedies than to either literature or life. There are just as many different kinds of old women as young women—why not? We don't really decline into types as we age; we are seen that way only because it is convenient to shuffle off the old and—along with them—belief in our own old age.

The strong old lady in Tibor Déry's stunning story "Two Women" incarnates an entire past—Vienna and Budapest in the days of the Austro-Hungarian Empire, the old-fashioned German humanistic culture, a life of ease and flirtation—and it is to this past that her daughter-in-law clings in order to survive life in Communist Hungary after the arrest of her husband. Concentrating on the moving rela-

tionship between the two women, Déry is still able to give both a brief but telling history of the decline of one way of life and a sense of the political oppression that shadows Lucy's performance in her mother-in-law's bedroom. The story is not gloomy, however, but glows with the achieved splendor of classical art.

TIBOR DÉRY

Two Women

I

In the morning, before she pressed the bell of her mother-in-law's small cottage, Lucy stopped when she reached the front door, took the small round mirror out of her handbag, and checked whether her tears had smeared her black mascara. The mirror glinted in the sun, casting a dazzling beam of light which flashed back from her shining red hair. She scrutinized her face with narrowed eyes, with the passionate objectivity of a tracker dog; to be on the safe side, she put on some powder, and retraced the line of her lips with two swift light strokes; she would have resented even Irene, the old housekeeper, seeing her with a tear-stained face. From the open kitchen window came a smell of buttered toast, and this made her feel a little better. Still standing in front of the door, she "concentrated" for a few seconds more, then, with her customary gesture, scornfully shrugged her shoulders. Up till now—meaning the best part of a year—she had always managed to scrape through the mornings somehow.

To the left, next to the three stone steps of the entrance, the spirea already stood in bloom, spilling its sparkling white flowers like a cascading waterfall onto the narrow strip of lawn. Beyond the fence, four women stood gossiping in the high road, and about a hundred yards farther on, four geese stood motionless in the dust.

"The old lady got out of bed with her left foot today," said Irene.

Lucy planted a kiss on the red, round face of the old housekeeper.

"Did she get up?" she exclaimed, laughing. "Has Christ come down and performed a miracle?"

The apple-cheeked housekeeper burst out laughing. The old lady, Lucy's ninety-six-year-old mother-in-law, had been laid low seven years ago with a kind of neuralgia; since then she got out of bed less and less frequently, and never before six in the afternoon, when—in a ratio directly corresponding to her waning strength—she walked the length of the room at first fifty times, then, slowly decreasing the daily quota, only about ten to fifteen times. She reduced it to ten in the last year, but had not fallen below that figure yet.

"She's been fuming ever since last night about not getting a new cover for her bedpan. Have you brought any savory biscuits?"

"Tons," said the young woman. "I've brought you some money for the housekeeping, too, Irene."

"Tons?"

Lucy laughed again; her laughter was irresistible, you couldn't help laughing with her. "Hell," she said. "I've brought you five hundred for the moment. Will that do?"

"That's more than we want this month," said the housekeeper.

The young woman opened her white sacklike handbag, and started rummaging in it. It was not tidy; the few folded banknotes she had received this morning at the Commission Store for her old blue-fox stole lay crumpled at the very bottom, beneath the savory biscuits, the little tray cloth wrapped in tissue paper, under her lipstick, her purse and her mirror.

"One-two-three-four-five . . ." she counted, and her finely cut, mobile face, whose expressive features never quite came to rest, looked for a moment deeply serious while she counted the money. "Never mind, we'll settle it if there's any left. You don't have to stint yourself, old thing," she said, and was already laughing again. "Any other trouble?"

"None. Got any cash left for yourself?"

"To stuff pigs with," the young woman asserted. Into the clean white kitchen where the gleaming whiteness was doubly increased by the glaring sunlight, the acacia, standing in front of the window, sent a bouquet of chokingly sweet fragrance that for a moment suppressed the aroma of hot buttered toast.

"Has the postman been?" the young woman asked.

"Not yet."

"Keep an eye on the letterbox, Irene dear; there will be a letter from America today," said Lucy. "I'd like her to read it while I'm here."

"Oh, she's ringing again," the housekeeper said. There were three or four short, impatient buzzes. "If she's rung for me once since breakfast, she's rung ten times. Her hand works as if it had an itch in it."

"Leave it, dear, I'll go," said Lucy.

Standing in the doorway—the opening creak of which did not register with the slightly deaf old lady—Lucy contemplated the thin, pallid face that was framed by a black velvet bonnet tied with a ribbon under the chin. The head had sunk into the huge white pillows, the large, straight waxen nose protruding from the bony structure of the face, beneath the closed eyelids, seemed to foreshadow the death mask. Not a muscle moved under the light yellow eiderdown, not even the rising and falling of the chest was visible. The young woman had to "concentrate" hard again; her eyes were begining to fill with tears. Fragile and athletic, she stood on the threshold, her head thrown back on the lovely, long stem of her neck as if the weight of her glorious russet hair were pulling it down toward the ground; her nostrils quivered, her small round belly tensed with excitement. *"Silly goose!"* she said to herself furiously in an effort to gain some strength once again. The sun shone from behind the enormous bed, on the narrow white face lost between the billowing pillows, that was saved from extinction solely by the black velvet bonnet. The young woman pulled a wry face and impatiently stamped her foot.

The woman lying on the bed opened her eyes and looked toward the door.

"Good morning, Mother," Lucy called, her voice ringing light and gay once more.

"I rang so often, Irene," the old woman complained huskily, and, knitting her brows, she looked toward the door again. "Is the buttered toast not ready yet?"

"It's me, Mother," Lucy called. "It will be ready in a minute."

The old woman cast another suspicious glance toward the door. "Who is?" she asked.

Although she had lived in Hungary for more than seventy years, the Austrian accent of her home town, Vienna, had not yet worn off. She spoke Hungarian fluently, but with a Germanic turn of phrase. A Hungarian word misheard and stubbornly retained by her mind in the crooked way it registered it when she first arrived occasionally slipped into her vocabulary, and each time this irked her daughter-in-law to uncontrollable laughter. She still read German in preference to Hungarian, and in these last years that she had spent in bed, she had returned to the adored ideals of her youth; she read Goethe, Schiller, and sometimes—with slight boredom—a volume of Thomas Mann and, out of consideration to the country which offered her hospitality, the essays of the Hungarian Gyorgy Lukács on Goethe; these she annotated in her large, masculine Gothic handwriting with innumerable marginal notes and exclamation marks. Her own remarks—mainly quotations from the books read, which were interspersed with household accounts—she penned in a hardbacked ruled copybook. The books—eight or ten volumes of them—the copybooks, one packet of sweet and one packet of savory biscuits, as well as a box of chocolates of which she consumed one per day, were ranged in precise and predetermined order at the edge of the large bed, so that they were easily to hand, and so that it should not be necessary for her to ring for Irene if she wanted any of them. The radio stood at the foot of the bed, tuned in to the Kossuth station, so that with the aid of a switch fixed to a long flex, put up for her by her son, she could operate it herself, and only had to call for the housekeeper when—according to the premarked program—the Petöfi station supplied the composition required, and the set had to be switched over. But for the last few months this had happened less and less frequently; the old lady lay prone in her bed for the best part of the day with closed eyes, and it was impossible to tell whether she was dozing, turning over the memories of her youth in her still agile brain, or speculating on her all too brief future. She came to life only when Irene offered her some forbidden paprika potatoes from her own lunch, or when her daughter-in-law called in the morning.

"You—is it?" she said, grumbling, slightly lifting her head from the topmost lilac silk pillow. "Haf you bring me the little salt cakes?"

"Of course I have," laughed Lucy, and, opening her sacklike handbag, she cautiously lifted out the little tissue-clad parcel. "And what else have I brought for you, Mother?"

"For surely, flowers," the old woman said discontentedly; her features, however, smoothed out instantly and her eyes, her still lovely dark eyes, lit up with delight when they fell on the small bunch of gold pansies. "You will never learn to economize?"

Lucy stepped right in front of the bed, rolled her eyes up so that only the white showed, and, extending her forefinger, began to admonish the old woman with it. She kept it up until the latter started laughing.

"All right," she said. "I say the same thing every day because you makes every day the same thing. You not to bring me flowers every day. You to buy me tomorrow eau-de-cologne."

"And what else have I brought for Mother?" Lucy asked, diving into her handbag again. She poked around in it for a long while, watching the old woman intensely in the meantime. On the old face sudden senility appeared, the moronic reflection of credulous, openmouthed infantile curiosity. Out of the dark cavity of the mouth glinted her one remaining tooth.

"Well, give it here," she said, losing her patience after a while. "Or have you got it lost?"

When she saw the little embroidered tray cloth Lucy took out of the tissue paper she flushed with pleasure. She was demanding, but at the same time childishly grateful for the smallest service. "Is good," she said, "you now to bend down and to kiss me." She slightly lifted her head and, pursing her bloodless lips, clasped one of her bony hands on the full white neck of the young woman, drew her down and kissed her forehead noisily. "It's beautiful," she said. "But where from you take so much money?"

"I earn it, Mother," Lucy said cheerfully. "You know what a prodigious whore I am."

The old woman looked at her with round, uncomprehending eyes for a while, then suddenly started giggling, and lifted a threatening finger.

"Whore?" she said. "Ach, go away. I shall tell you to Janos when he comes home, and you will get two such big boxes on the ear." She turned her black-bonneted head toward the bedside table, and looked at the bedpan standing on the lower shelf. "You to cover it up," she said. "It is lovely. It is high time, too, because in the afternoon comes

Professor Hetenyi, and Irene will get excited and maybe will forget to put it away like the last time, and then here we are."

"The Professor's seen worse than that before," said Irene, appearing suddenly in the open doorway. "And anyway, what makes you think he'll come this afternoon?"

Wrinkling her forehead, the old lady looked toward the door. "She always eavesdrops," she whispered reproachfully to her daughter-in-law, but as her deaf ear could no longer censure the volume of her voice, her subdued hiss sounded as if she had bawled the old housekeeper out.

"Of course she wasn't eavesdropping," said Lucy quickly. "She's brought you the buttered toast."

"That I should be eavesdropping . . ."

Lucy winked at the old housekeeper like a conspirator.

"That I'm eavesdropping . . ."

Through the open window came a loud cackling of geese.

"That *I'm* eavesdropping . . ."

Lucy burst out laughing. "Don't have a stroke, Irene dear. If you go on like this I'm sure you will. Come on, let's have that toast."

"That is why the Professor comes today afternoon," said the old lady from the bed, angrily, toward the door, "because he was not here already for very long, for at least a month, and he has promised that he will come once every month."

"He is coming this afternoon," said Lucy teasing, "because Mother is in love with him, and is expecting him every afternoon; so that she can chat with him about Goethe—and she doesn't want the Professor to look at her bedpan while they're talking, why, that's only natural, Irene dear, isn't it?"

"What do you say?" asked the old woman, blushing. "What do you say?" When it suited her, the old lady could go stone deaf in a second. "Why must you always so whisper so that one cannot hear one's own voice?" Lucy laughed outright, and the geese in front of the house started cackling again. "What do you say?"

The reason Professor Hetenyi had not been to the little cottage for so long was that he had died four months ago, but this was a secret closely guarded from the old woman, who was not able to keep count of the passage of earthly time any longer.

"It's no use denying it, Mother," said Lucy, "that you are in love with the Professor. Otherwise, for whom else would you learn all those quotations from Goethe by heart?"

The old woman had in the meantime fitted her dentures into her mouth, and was now crunching the toast. *"Dumme Gans,"* she said. "For whom? For whom? For myself. I'm eating the toast for myself, too."

Lucy sat down in the little armchair next to the bed, that was upholstered in yellow silk. She crossed her legs. "Tell me, Mother," she said, "is the Professor really so witty when he speaks German?"

The old lady did not reply, but continued to munch her toast disdainfully.

"Well, he might be witty when he is talking German," went on Lucy. "Actually, he's quite sweet even when he chatters away in Hungarian. If only he weren't so ugly."

The old lady still did not notice that she was having her leg pulled. "Not everybody can be as beautiful as you," she said, and a slight flush spread over her forehead.

"Am I beautiful?" Lucy asked.

The old woman stopped pushing the food around in her mouth, and her old eyes, seasoned with the experience of nine decades, fastened attentively on her daughter-in-law's face, every sensual particle of which was bathed at this moment in the sunlight, and scent of acacias streaming in from the window.

"If only were you so clever," she said querulously after a time.

"As clever as whom?"

The old lady started rotating her food again and did not answer.

"As clever as the Professor?"

"Dumme Gans," said the old woman.

The young woman laughed cheerfully. "Oh, I nearly forgot, the Professor rang yesterday," she said. "He only came up for the day from Szeged, so he couldn't come to see Mother, but he sends his respects. He says Mother will live for a hundred years, so not to worry, there will be many more times when they can see one another."

The old lady went on eating, but cast a suspicious glance at Lucy from the corner of her eye. "He told me too that I will live one hundred years," she muttered after a while. "Did he now say the same on the telephone to you?"

Resting her head on the sun-drenched back of the armchair, the young woman closed her eyes.

"You not now making a fool of me?", the old woman inquired.

Smiling with closed eyes, Lucy nodded.

"How? . . . What? You now a tricking me?" the old woman asked incredulously.

"Of course," said Lucy.

The old lady looked at her with large naïve eyes. Now she really did not know what to make of the Professor's prediction. But whenever Lucy laughed, then, however put out she was, she ended up laughing with her daughter-in-law, with quiet, perplexed chuckles at first, her bony austere face assuming an expression of fatuous simplicity, then more and more relaxedly, with more honesty and with the good-humored realization that she was laughing at her own expense—the tiny sting left in her brain stabbing her memory only hours later and sometimes only the next day.

"You are now really fibbing me?" she asserted gaily.

Lucy jumped up, bent over the bed, and with her fresh young mouth planted kisses all over the hard old face trembling between the pillows.

"How can you think that I'm telling you a fib?" she exclaimed. "Haven't we worked it out a few days ago who is going to be still alive on Mother's hundredth birthday, and whom we can invite, and even what dress I am going to wear? We said my low-cut red silk one, don't you remember?"

"We haf not choose the white silk with the low cut?"

"Where is that by now?" sighed Lucy quietly.

"What have you say?"

"Nothing, Mother."

"All right," said the old lady. "Then you give me soda water now. Has to you the Professor really said on the telephone that I will live one hundred year?"

While Lucy was rearranging the crumpled pillows in the bed, and the old lady cautiously groaned from time to time from the remembrance of the neuralgia of seven years ago, Irene came into the room again, and, with a jealous glance at the quick skillful hands of the young woman, removed the little wooden tray with the empty glass and plate on it.

"Aren't all these books a nuisance for you, Mother?" asked Lucy, smoothing the eiderdown. "You can only read one at a time!"

"But I am reflecting on the others," the old woman said severely. "You not to touch those ones either. Have you learn the German lesson?"

As she did not receive any reply for quite a while, she turned her narrow head toward the young woman, fixing her with a reproachful gaze. The gaze became a stare, losing its focus and purpose, and lasted for so long that it seemed to have lost all relation to subject and object alike, and was now glimmering on its own with aimless radiance, like an orb in the ether. Then, suddenly, it became affiliated with the earthly world again.

"What is Janos going to say," she said accusingly, "when he comes home and you not speak German yet?"

"Janos. . . . Janos. . . ." Lucy mumbled. "He'll be lucky if I haven't forgotten Hungarian by then."

The old woman looked at her again. "What do you say?"

"I'm tired of Janos," muttered Lucy, but of course the old woman caught that. "I'm tired of Janos," Lucy repeated slightly louder to make sure she would be heard. "How is it that whenever one mentions Janos you always hear it, however quietly one speaks, Mother? I'm sick of your son, Mother."

"You are now fibbing me again?" the old woman asked, her mouth slightly twisted by laughter.

"Of course," said Lucy.

The old lady didn't know what to say. But her eyes, resting on the face of her daughter-in-law, were so anguished and so guileless that Lucy could not bear it for long; she raised her head and smiled. Immediately and gratefully the old woman smiled back.

"Tell me, Mother," Lucy inquired, "how was that when once you were staying at Savanyukut?"

The old woman reflected, "Savanyukut . . ."

"Yes," said Lucy, "you know, when the whole family had to sleep in one room."

"And?"

"Oh, you know," said Lucy, "when you all had to sleep in one room and your other son was peeping."

"Poor Gyuri? You did not know him any more."

"Yeah," said Lucy, "I mean Gyuri. When in the evening you had to get undressed, Mother, and told the children to turn toward the wall."

"Ach," said the old woman. "I haf told you about that before."

Lucy smiled at her again. "But I don't remember it very well. When was it?"

"Ach, very long ago," the old lady sighed. "Maybe it is fifty years."

"More than that," said Lucy. "How old are you now, Mother?"

"Ninety-six," said the old lady proudly.

"There you are!" Lucy exclaimed triumphantly. "And when you were at Savanyukut, and could only get a single room for the night, you were still quite a young woman, weren't you? That must have been at least sixty years ago."

"Sixty?" the old woman mused. "Then was I thirty-six?"

"Just about," said Lucy. "Now tell me, how was it with that spying?"

"Well, what happened . . ." the old lady started, then stopped abruptly, and shot an uncertain glance at her daughter-in-law. "But I haf told you that already."

"But if I don't remember it very well," complained Lucy. "I only remember that it was evening and the children were already in bed, and that you wanted to undress. Mother, you must have been very beautiful as a young woman."

"Yes, I was very beautiful," said the old woman simply. "You know, at that time became fashionable those big hats that you had to tie down with those silk or velvet ribbons—how do you say, under your chinbone, and my husband, he gave me as a present such a very wide moiré ribbon, and when I tied that down under my chin every young man turned around after me in the street."

"I'm sure they did," said Lucy, "and not only the young ones either. Professor Hetenyi would have turned around, too, if he had known Mother at that time."

The old lady smiled modestly. "Oh, go away," she said, suddenly sternly, "maybe the Professor was not even alive then."

"Maybe," said Lucy, "anyway, what happened then?"

"When?"

"Well, about the spying?"

The old woman ran her bony finger over her black bonnet. Do you know," she said reminiscently, "that this bonnet has been made from that very wide ribbon that my husband gave me then?"

"But this is velvet, Mother!" Lucy laughed merrily. "And you received a silk moiré ribbon."

"Never mind," said the old woman, flushing. "Maybe that one was velvet. Well, what you want to know?"

"Well, Mother," Lucy said, "what I would like to know is how you undressed that time at Savanyukut."

"Everybody had to sleep in one room," the old woman began, and her voice sounded slightly muffled from crossing the crippling distance of sixty years in the dimension of time; even her eyes lost some of their luster. "What happened was that everybody had to sleep in one room and the children were already lying in their beds, and I wanted myself to undress. Do you know, all the ladies wore corsets in those days, long corsets from here to here, and that my man had to lace it up, and that I haf not wanted the children to see. Oh," she said impatiently, "I haf told you all this before."

"But if I don't remember it, Mother," Lucy said. "All I remember is that you told the children to turn towards the wall."

The old lady nodded distractedly. "I told them."

"And when you started to undress, you looked behind you to make sure that the children really had turned to the wall."

"Yes," said the old lady, "I stood there in my petticoat."

"Sorry, not in a petticoat," interrupted Lucy, with lips twitching with laughter, "but in pants, in long pants, down to your ankles, with lace at the bottom."

The old lady laughed, too. "Yes, in pants."

"And when you turned to look, you saw that Janos was honorably turned toward the wall, but your younger son . . ."

The old lady burst into laughter again. "Yes, Gyuri was espionaging."

"Yes, he was espionaging," said Lucy. "He was lying on his tummy and he was ogling Mother with one eye."

"He had long golden hair," the old woman said softly, "lovely, long golden locks, and he was espionaging from behind them into the room."

"And then?"

"What then?"

"Then what happened?"

The old lady did not reply.

"Oh, do go on, Mother," said Lucy, "because I don't remember all of it. You put a large towel over the side of the crib . . ."

"Of course, I did put one over it," the old woman said vaguely.

Lucy giggled. "And then you saw from the mirror that Gyuri was lifting the corner of the towel and peeping from behind it."

"Yes, he was peeping," said the old lady, and in order to be able to laugh more comfortably, she took her dentures out of her mouth and placed them in the glass of water standing on her bedside table. "He was peeping out and grinning off his head."

"While Janos . . ."

The old woman suddenly glowed with the memory. "While Janos," she said quietly, "he just lay there, turned toward the wall, and did not move."

"He was very scrupulous even as a little boy, wasn't he, Mother?" asked Lucy, watching the old woman closely.

"Yes," said the latter, very slowly and decisively. "He is a very honorable man."

She closed her eyes and there was silence in the sunlit room, broken eventually by the deep hooting of a horn that came from outside the window. The old lady opened her eyes again, and fixed them on Lucy. *"Dumme Gans,"* she said loudly and deliberately, "you belief that I do not see that you remember better than I? Yes, he was all that scrupulous even as a child—is that what you wanted to hear?"

"Like hell I did," Lucy said quietly, but the old lady did not hear it. She was watching a wasp that was circling above the little bunch of pansies that stood on the bedside table; craning her neck, she was watching it with her head turned sideways, then—forgetting her neuralgia—she propped herself up on her elbow, and, popping her spectacles on her nose, her mouth half opened with rapt attention.

"You love Janos?" she asked hoarsely, after a while. Exhausted, she sank back on her pillow and took her glasses off.

"Like hell I do," said Lucy.

The old lady knitted her brows with disapproval. "All right, you not to fib me now," she said huskily, "because I now was speaking in earnest. If you love Janos you tell him when he comes back from America that he should not be so obstinate."

"Why, Mother?" Lucy asked with interest.

The old lady did not answer. She was staring into space, her mouth half opening again with inward excitement. "Never mind," she said. "You just tell him that."

"But why, Mother?" asked Lucy again.

Again the old woman made no reply. She lifted her head a little, tried to include in her vision the wasp droning above the pansies, observed it a little while, and then let her head fall back on to her lilac pillow.

"Once I was watching such a wasp," she said in a muffled voice, "that was buzzing too like this one above such yellow pansies, then sat down on one and because it was very heavy the pansy quickly turned her head and bent deep to the ground. You know, if it had not bent down, then perhaps its . . . what you call it—stem would have broke."

"Is that what I am to tell Janos?" asked Lucy.

"Yes," said the old lady.

"It's too late," muttered Lucy.

"What do you say?"

"I haven't said anything."

"You don't have to," said the old lady. "You just tell him that."

"Why don't you say it to him, Mother?"

The old lady lifted a hand, and let it fall wearily. "It will be a very long time before he comes home," she said, staring fixedly into space again.

Fortunately Irene came into the room again; her pleasant round face shone as if clad in its Sunday best; even her white hair sparkled in the sunlight. The old lady spied the letter she was holding at once. She sat up, opened her mouth, but no words came.

"There is a letter for you," said Irene, winking at Lucy.

The old lady took a deep breath. "From Janos?"

"How should I know?" said Irene, coming up to the bed. "I haven't read it."

But this time the old lady was not to be taken in by the indifference of the tone; as if an earthquake had shaken and transformed her whole being within the space of a second, her eyes dilated, and the cavities of her long bony face filled out, color crept into her cheeks, and her breast was heaving under the light yellow eiderdown. The weight of the excitement brought her soul up to the surface, into the pupils of her eyes, her colorless nails, her scanty, electrified white hair. She extended both her trembling hands; with one she grabbed the letter, with the other she clasped the old housekeeper's neck, pulled her head down and kissed her. She felt so faint that it was minutes before she could open the envelope, tear the letter out of it, and smooth out the typewritten pages. "Give here the magnifying glass, Irene," she said, panting.

She put her spectacles on, and, holding the magnifying glass in her shaky hand, stopping from time to time as if she was out of breath, then, with a deep sigh, starting again, she read the four closely typed pages which for the next week she was going to take out every day and read, from the first letter to the last, over and over again.

This time I'll try to make up for my short letter, although I have even less time than before because—and this is the great news—I'm just putting the finishing touches to my film, in another month it will be ready and then I shall come home. This is the news I wanted to tell you first to compensate for the long wait you had. The first night of the film will be a month from today in a New York film theater which holds thirty thousand people. It is just being built on the outskirts of the city on top of a high mountain and has to be completed for the first night because they want to open it with my film. From the roof you can see half of America down to the Cordilleras and the Andes, not to speak of the Atlantic Ocean, which is just as blue here as the Adriatic at Abbazia where we were together one summer. The theater is going to have an airport of its own because all the rich people from all over the States will be coming here, flying their own airplanes, and from the city, meaning New York, there will be a special helicopter service for the audience, costing one dollar per head. They are preparing great festivities for the first night, but I'll come back to that later; first I want to reassure you that in spite of all the work I'm in the best of health and very happy about being able to return home soon, and although I work ten to twelve hours a day I reserve one hour every day for sport because I don't want to get fat. I box, fly a plane, walk and swim. Of course, you will ask, where do I swim? In my own swimming pool in the park. That's a story for itself.

You know, Mother, that until recently I stayed in the largest hotel of the United States, the Waldorf-Astoria; I had a suite of twenty rooms on the hundredth floor from which again you could see about half of America. But I had to keep so big a staff, about six secretaries and a dozen typists and so on, that there wasn't enough room for us all although in the end I had beds put up even in the reception rooms, in bunks like the ones you saw on your honeymoon! However, the reporters who kept coming from all parts of the world pestered me so much that we decided to move out and keep our address secret. So, about a thousand miles from New York, we rented a sixteenth-century castle, every stone of which was imported from France separately, and rebuilt here, of course with every modern comfort, in the middle of a hundred-acre park that has several swimming pools and an airport as well. The house is air-conditioned, and now we have enough space because the castle has so many rooms that even my black valet and my hairdresser have a suite of their own, and my groom a small cottage next to the stable. Oh, I forgot to tell you, Mother, that I go for an hour's ride every morning before breakfast, I have a lovely white mare called Darling, and I canter away on her, leaving even my secretaries behind. At these times I'm followed only by the six secret police officers detailed for my safety by the film company, but they ride a hundred paces behind me. If we should have anything to do in New York we can fly there in less than an hour in our own red-white-green jet plane, and can get back before lunch, because, as you know, Mother, I insist on my afternoon siesta, so I can be fresh and rested for work in the evening. *My address is secret, so I can't tell even you, Mother, please give your letters to Lucy as before, and she will send them to my old address, the Waldorf.*

All these distinctions I've been overwhelmed with are most embarrassing; I only write about them because I know they please you, Mother, and this will make up to a small degree for the distance that separates us. Still, this will come to an end soon, and I shall be home again. Unfortunately, I'm unable to send anything of real value to you, Mother, for the moment, as I shall only receive my salary after the film is finished, and anyway, you would have to pay too much duty at home. So, for the time being, you will have to content yourself with the bits and pieces I can send through Lucy. But I have already earmarked a large silver-gray luxury Ford for you; I shall bring it home myself, as well as a phonograph with a thousand records and a superb radio on which you can even get the North Pole. (Although they don't play Beethoven there!) For Irene I'm going to buy a little five- or six-roomed villa somewhere in the neighborhood, but I'm not telling anyone what I have for Lucy. There is a large brown bear here in the garden, about twice as big as Irene; we are very fond of each other, so maybe they'll sell him to me—although I have no idea where we could keep him at home. He looks rather like that big bear we saw together in Vienna when you took me to the Zoo in Schönbrunn when I was ten. Do you remember? And the same evening we went to see Offenbach's *Helen of Troy,* and I laughed so loud during the performance out of sheer joy that the whole audience was looking at me—do you remember that too?

Oh, I still owe you the description of the festivities they are planning for the first night. Don't speak too much about them, because they are partly a diplomatic secret. I don't suppose I need to tell you that all the radio and television companies of the States are going to arrange for a live transmission so that at least fifty million people will see the film. They are expecting one or two hundred reporters, and the best-known radio commentators, among them the world-famous Mr. Smith, and of course the representatives of the European press and radio companies are going to be present too. Pity Budapest won't be able to receive the transmissions.

They have invited a lot of celebrities in my honor. I'm saying that just to please you, Mother, not to boast. Mrs. Roosevelt, the widow of the late President, has undertaken the chairmanship of the festivities, and invited her friend Queen Wilhelmina of the Netherlands, who has accepted the invitation. The President of the French Republic is also coming with his whole family, and representing the Queen of England will be her son, the Prince of Wales, whom you, Mother, will no doubt remember. He was in Budapest once and rather liked our apricot brandy. The foreign minister of the Soviet Union is coming, and Hungary will be represented by the Minister of Culture; the German Chancellor and the King of Greece are coming—the only thing they are wondering about now is whether or not to invite Franco. The Maharaja of Jaipur has accepted, and is bringing his whole court with him, also his favorite white elephant, who of course cannot be dragged up the mountain, so they will have to build him a separate stable at the foot of it, all of white marble and air-conditioned, and also some cottages for the elephant-keepers. The elephant has a pure gold saddle studded with pearls and precious stones. . . .

By the time the old lady finished the letter, reaction to her excitement set in, and she fell asleep quite suddenly. It was possible to talk freely now, for if she did not see lips moving, she did not register voices, even if they spoke next to her ear. She slept quietly, without a sound, her bony face transformed by the magic in happiness. While

she was still on the whole in full possession of her senses, and occasionally—as far as her naïve disposition permitted—even capable of
suspicion, she was quite prepared to believe the most fantastic things
of her son; his talent, his strength, his courage, his genius, all were
beyond question.

Irene giggled. "Again she didn't notice that the letter was posted
in Budapest," she said.

Lucy jumped up from her chair, went to the window, and from
behind the bed, so that she should not even see it in her dream, stuck
out her tongue at the old woman.

"What did you write?"

"A lot," said Lucy. "I laid it on thick."

"She won't talk about it today," said the housekeeper, "but tomorrow she'll ring for me at dawn."

"Well, why shouldn't you have some fun, too?"

They were both silent. The sun climbing to its zenith had left the
room, the old lady's face was now in shadow. Only the mirror of the
large wardrobe standing at the far end of the room reflected the sun's
rays, the beveled edges dissecting it to the seven colors of the rainbow, and throwing a trembling veil of colored light onto the wall.
Above the bunch of pansies the wasps hummed sleepily.

"You won't be able to stand this much longer," said Irene, looking
with compassion at the young woman.

"Like hell I won't," said Lucy.

"Well—not for ten years."

"He's done one," said Lucy, "there are only nine left."

Irene nodded. "The old lady won't live that long."

"No," said Lucy, "but as long as she lives, I want her to have
everything."

"Professor Hetenyi promised her she'll live to be a hundred. That's
four more years."

"Possibly," said Lucy. "Come over here, Irene dear, and have a look
how lovely her face still is from this angle."

The old woman slept so soundly and so happily that Lucy decided
she would not wait for her to wake up. But before she had reached the
door, the old lady behind her opened her eyes. "Where are you going?"

Lucy turned around, laughing. "I was going to hop it, Mother."

The old lady said nothing; she looked at the flexible, slim figure of her daughter-in-law for a long time. She lifted her head a little, and smiled gently.

"How lovely your red hair looks in the sun," she said. "Like an English princess."

Lucy flushed. "Well! Is it really lovely, Mother?"

"Really," the old lady said. She nodded a few more times, then called Lucy to her. "Will you kiss me, please? No, not there, on my forehead. I want you to. You are very lovely and very good."

"Well, really!" said Lucy.

"There's something I want to ask," the old lady went on. "Why must my son Janos have secret police by him in America?"

Lucy swallowed hard; her eyelids began to flutter.

"Well, every famous person is surrounded by secret police in America, Mother."

"All right," said the old woman, relieved. "Is good." Suddenly she laughed in a thin old voice. "Do you want one bear? One very big bear so big as two Irenes. *Mein Gott,* what an idea—only he can have an idea like that." She laughed until the tears came to her eyes. "Now you go," she said, suddenly impatient. "Tomorrow I will tell you the letter."

But Lucy had hardly reached the door before she called her back.

"Now," she said, "you nearly forgot to pull out the hairs on my chin, and this afternoon Professor Hetenyi is coming. You please take the tweezers out from my drawer. And the new bedjacket, because Irene will not give it to me. You to put it on me, please. You see," she said proudly, while looking at her bare arm, "you see how it is nicely smooth, there is no wrinkle in it only where I bend, here by the elbow."

"The young ones have wrinkles there, too, Mother," said Lucy.

The old woman looked at her.

II

That morning Lucy arrived at her mother-in-law's cottage later than usual. Opposite the door four geese were cackling in the sun-flecked sand—the four women already had gone home to cook the mid-day meal.

The door was opened by the white-haired, square-set housekeeper. She did not allow the visitor to enter at once; she planted herself in front of Lucy, and with her small sharp eyes set in her round apple face she scrutinized the young woman's features for a long time.

"Well, aren't you going to let me in?" Lucy asked, laughing.

"What's wrong?" asked Irene, blocking the doorway.

"Nothing," said Lucy. "Why should there be anything wrong? Aren't you going to let me in?"

The old housekeeper did not move from the doorway.

"There is nothing wrong—why should there be?" Lucy repeated.

"Why didn't you bring any flowers, then?" she asked.

Lucy laughed. "Aren't you sharp! I didn't have time. Well, let me in, blow it. I'm late as it is."

"You didn't have time?"

In the clean white kitchen, where the gleaming whiteness of the furniture was enhanced by the blazing sunlight, the jasmine, from the edge of the lawn, sent a bouquet of chokingly sweet fragrance. Turning her head from side to side, Lucy sniffed the air. "I can't smell any buttered toast," she said. "Has she eaten it by now?"

"Ages ago."

"Is it that late?"

"Nearly midday," the housekeeper said.

"Irene dear, go and pinch yesterday's flowers out of the vase," said Lucy, "and bring them out here. I had no time to buy any today."

"Are you broke?" the housekeeper asked.

Lucy did not reply. "Look, just pinch yesterday's flowers," she said, "and give me some tissue paper to wrap around the stems. What sort of flowers did I bring yesterday?"

The housekeeper shook her head. "I can't pinch them because if she never notices anything else, she'll notice if the flowers are missing, sure enough. You brought white carnations."

"Bring them out," said Lucy. "I'll put some lipstick on them."

Irene laughed. "Would you like a glass of brandy?"

"Yes."

Irene went to the kitchen cabinet and opened it. On the lower shelf of the top part, which was edged with a white doily strip, the soup bowls, dinner plates and dessert plates shone in separate groups, the last group a little lower than the rest. The meat, vegetable, salad and sweet dishes were piled upon one another according to size, the great urn-shaped soup tureen was in the far corner. They were all decorated with a narrow gold band which glinted in the sun.

"She broke another dessert plate yesterday," the housekeeper said, while from among the glasses, cups and saucers on the top shelf she took a lead-crystal liqueur glass and filled it with brandy.

"What's the matter?" she asked as, turning round with the full glass in her hand, she saw that Lucy, her head resting on her arms on the kitchen table, was crying soundlessly. "I knew . . ."

Annoyed, the young woman threw her head back. "What did you know? Tell me, what did you know?"

The old housekeeper watched her silently.

"Well, is it surprising if I cry," sniffed Lucy, "if they break my inheritance to pieces? That's the fourth dessert plate that's missing. Or did she leave the china to you?"

"She did," said the housekeeper quickly and decidedly. "And the silver too. What's wrong?"

"Nothing."

"No money?"

"I made seventy-four forints last week by dyeing nylons," she said. From her white handbag she took her handkerchief, powder, lipstick and mirror, and threw them down on the table in front of her. There were a few specks of powder on the mirror, which she wiped off with her handkerchief, and then raised it to her face; with the heartbreaking intensity of the eternal struggle for existence she examined each feature with infinite care. Moving the mirror from side to side, she smoothed her greasy eyelids with the tip of her forefinger, sucked her lips in and pressed the upper lip to the lower a few times. Then, wetting her forefinger, she passed it once or twice over her eyebrows. Creation, without which there is neither life nor death, had been completed. Looking into the mirror once more, she cast a last glance at her heavy russet hair; this was Sunday.

"I only had to work two days and two nights for it," she said, the last tremors of creation reverberating in her voice.

The housekeeper pulled up the little kitchen stool where, under the worn paint, the texture of the wood was beginning to show through, and sat down next to her. "I could lend you some," she said. "My husband got his wages yesterday."

"Go to bloody . . ." said Lucy. "The hairdresser offered me eight hundred forints for my hair only last week—that should do for two months' rent as it is."

"Is there no news?"

"None. My mother-in-law?"

"She was asleep when I looked in a few minutes ago."

Lucy's eyes filled with tears again. She could not even put up any resistance to herself any more—she was finished.

"What's the matter with you?" Irene asked.

"Nothing. I'm jittery. Leave me alone."

"Got your period?"

Lucy laughed suddenly. "Sure. That's the least of my worries."

"Well, what then?"

"The bailiff has been again," the young woman said. "Came at nine and plagued me till eleven."

"But he's already collared everything."

"No."

"What d'you mean?"

"I mean he hasn't confiscated everything."

The old housekeeper folded her hands in her lap. The geese broke into a frightened cackle in front of the house.

"As soon as he left," went on Lucy, "the family of the concierge came up, all three of them, father, mother and the little girl, and sang Verdi's *Requiem* for me."

"What?"

"Nothing. I'm playing the fool," said Lucy.

The housekeeper stood up, then sat down again.

"When I was a little girl," said Lucy, knitting her brows, "I once got lice at the convent in Majsa. The Mother Superior ordered my hair to be shorn off completely, and it took a year until it grew anyway near halfway respectable again. In that year I felt more or less as I feel now."

The housekeeper was silent.

"Funny," said Lucy, "I just remembered, the Mother Superior came from Vienna, too. And I was often hurt because she always spoke German with Lilla Bulyovsky and I didn't understand it."

"She made you bald? Cut off all your lovely red hair?"

"She cut it off because it was red," Lucy said. "And the next day my mother went to the convent with a whip, and had to be forcibly restrained from horsewhipping the Mother Superior."

"I thought the bailiff had laid his filthy hand on everything by now?" the housekeeper said after a while.

"He hasn't," said the young woman obstinately.

"Well, what did he leave?"

Lucy made an impatient gesture and bumped her elbow on the table. "No!" she cried in fright, "I don't want an unexpected guest! I don't want any unexpected guests!" Snatching at her elbow with her hand, she pulled the unexpected guest out of it, and, turning her palm outward, scattered him into the four winds. "I've had quite enough of them as it is. He hasn't dispossessed me of my furniture yet."

"Of course not," said Irene.

The young woman shook her head with annoyance. "Why 'of course'?" she asked. "Because I had all the old bills and notes of delivery and with those I could prove conclusively that the furniture belongs to me and not to Janos. Janos had hardly enough furniture for one room—he didn't even have a decent bed when he married me."

"I know," nodded the housekeeper.

"And now the Ministry has sent Mr. Kulinka along to confiscate everything, and I'm supposed to reclaim whatever belongs to me," said Lucy with tears in her eyes. "He was just spitting noughts all over the place—five thousand, ten thousand, another ten thousand—it was no use begging him and telling him that they would restore the things to me anyway, and would he mind valuing them slightly lower, as I won't be able to cope with the stamp duty. He looked at me mournfully, for he was so sorry for me he nearly cried, then bawled out, 'One neo-baroque wardrobe, twenty thousand.'"

"Twenty thousand," repeated the housekeeper, shaking her head.

"What are you groaning about?" asked Lucy. "It's worth it. I'd get that much for it if I sold it any day. But I won't sell it, not if it kills me—I want Janos to find everything as he left it."

The housekeeper glued her eyes on the ground.

"Stop bawling," shouted Lucy irritably. "What the hell are you bawling about? Just like Mr. Kulinka. 'We won't frisk you personally,' he cried with tears absolutely flooding his eyes, 'but we'll search every cupboard.' . . . Irene dear, find me that old French soapbox we saved."

"Did you bring her some soap?" asked Irene, going toward the kitchen cabinet.

"Yeah . . . and find that empty Czech chocolate box for me, too, that nice red one," said Lucy, while taking a tablet of soap and a bag of candy from her handbag.

"She's still got the old ones untouched."

"Well, let her have some more," sneered Lucy. "Her beloved son Janos is sending them to her from America," she said sarcastically, while pouring the chocolates from the paper bag onto the table, and putting them one by one into the red candy box. She laughed again.

"What's up?" asked Irene.

"Nothing. I'm laughing at myself."

The bell buzzed three or four times in quick succession. The housekeeper looked at Lucy. "Are you going?"

"Not yet. I'll get my breath back first," said Lucy.

"Don't you want to lie down for a quarter of an hour in my room?"

"What for?" asked Lucy. "I'll lie down after lunch anyway, and won't even bother to get up till tomorrow morning. I'm out of work again."

"Why don't you go to the pictures for once?" Irene asked after a while.

"What for? I laugh enough about myself as it is."

The bell rang again. Since the time, about a month ago, when during one of her walks along the room, the old lady fainted, fell and broke her ankle, she had become even more neurotic and crotchety than before her accident. Sometimes she rang for Irene three or four times in quick succession; at other times she did not touch the bell for days; she lay in bed without moving, her foot in plaster, her eyes closed, neither reading nor switching on the radio; the only reason one knew she was awake was that she opened her eyes immediately some-one entered the room. At these times the mirror of her eyes was not clouded by the unconsciousness of sleep but by the weariness of a greater distance that drained the hue of life from her waxen skin and allowed the uncontrolled glands at the corner of her mouth to dis-charge their saliva. She did not speak at once at these times as dozing old people usually do who, like defenseless beasts, want to deny having been asleep, and do so with immediate and loud protestations; she inspected the fact that swam into her vision with a long and penetrating gaze as if trying to decide what strange unknown world it came from. She usually made some small deprecating gesture so as to wave away the unwelcome guest. Since breaking her ankle she could not get up in the afternoon, and, although she was not in pain, and was unaware that her ankle was broken—she thought it only sprained—the quarter of an hour taken from her daily program upset her whole routine and disrupted her small but steady universe. In place of the dead Professor Hetenyi—who was supposed to be furthering his studies in the Soviet Union—a new doctor had to be introduced to the scene, and this too contributed to her feeling of insecurity. Although she liked the new doctor, Dr. Illes, at sight as, for some unknown reason, through some unfathomable channels of the nervous system, he evoked the memory of her son Janos; the stranger intruding on her horizon still disturbed the groove her life had been running in for years now. Naturally, Dr. Illes spoke German well and had an extensive knowledge of the great German classics—they would not have dared to bring the old lady a doctor who did not—and apart from this he had a very nice tenor voice, and sometimes sang in concerts; this was something not even Professor Hetenyi could do. Equally naturally, Dr. Illes, following the example of his predecessor, shared, without any reservation, the old lady's admiration for her son.

"Is that you, Irene?" the old woman's voice came from the bedroom.

Lucy jumped up and tiptoed to the door to listen. She moved with the same grace when she was alone as when she was in company, even if only in her mother-in-law's company, who did not see very well now. Her face, complementing her body, acted as well—she grimaced, raised her eyebrows, wrinkled her forehead, smiled at herself as if to a second person who would watch, observe, and appreciate all this. Incidentally, her ears were so finely tuned that she could just as well have stayed in the kitchen, and would still have heard every word from upstairs.

"It is you, Irene?"

"Well, who else?" asked the housekeeper. "You rang."

"Come nearer, Irene dear, I do see your face not," said the old woman suspiciously. "I have rang?"

"Twice."

"I remember not," the old woman said. "Do you not know what I wanted?"

"I'm supposed to know that as well," muttered the housekeeper. "You wanted me to ask me what the time was."

"All right," said the old woman. "Tell me what is the time?"

"Nearly midday."

"Nearly midday," repeated the old woman pensively. "Did I not want anything else?"

"You wanted to ask me what I'm having for lunch."

"What are you having?"

"What's left over from yesterday's braised beef with potatoes and tomato sauce."

"Phew!" said the old woman. "I do not want that one. No noodles with cabbage?"

Irene shook her head.

"Pity," said the old woman disappointedly. "That Professor Hetenyi has strictly forbidden, but I always eat up by the spoonful. Tell me what time it is?"

"Midday."

"I know . . . midday," said the old woman. "But then why did not the young lady come yet?"

This was what Lucy wanted to hear; she grinned contentedly. It seemed the old lady wanted her after all, if for nothing else to talk about her son. She ran back to the kitchen, snatched up her handbag, had a last look in the mirror, and then ran back to the room.

"Good morning, Mother," she called loudly.

"Is it you?" said the old lady, bad-temperedly. "Why do you come so late? Irene says it's midday already."

"I wrote a long letter to Janos, Mother," said Lucy. "I took it out to the airport so it should go today."

"*Ja,*" said the old woman.

"What have I brought for Mother?" Lucy sang.

"Flowers, for sure," said the old woman. "You never learn to economize. What flowers they are?"

"White carnations."

The old woman turned her head toward the bedside table, where she could not see yesterday's white carnations. She looked back uncertainly to the vase in Lucy's hand.

"I put them in quickly with the others," Lucy said. "They wilted a bit on the long journey. And what else have I brought for Mother?"

she went on quickly, with distracting tactics. But the old woman did not look toward the handbag; she was still staring at the white carnations. "What else have I brought for Mother?" Lucy repeated more loudly.

The old lady looked at the presents with indifference. "Is good," she said. "French soap? Janos sent it? It smells very nice. You to give it Irene she will put it away. The candy too. You to write to Janos he not now send any more because there is now enough."

"Oh, let him send some," said Lucy. "There is never enough of these. Why shouldn't he send any?"

"Because he should economize," said the old woman wearily. "He cannot economize, either. What will happen to you, my child, if you will never learn to economize, not you, not he? One day you will be old."

"Never, Mother," Lucy laughed. "Not while there is lipstick and makeup on the market."

The old woman closed her eyes. "I thought it was already afternoon," she complained, "and that you already gone to home. And it is only midday and I still have to eat that miserable dinner."

Suddenly she sat up and, forgetting her neuralgia, leaned out of bed, fixing the dark rays of her dimming eyes on her daughter-in-law's face.

"Come nearer here, my child," she said.

"Yes, Mother."

"Nearer," said the old woman. "More nearer! Tell me, my child, the truth—when is he coming home?"

"Who?" asked Lucy, frightened.

It was the first time in a year that the old woman allowed her doubts to rise to the surface, and admitted openly even to herself that she had them. It had happened before that she had lost patience and grumbled that her son should leave her alone for so long, but she never doubted his veracity that in a month's time, or in three months, he would come home, and when that time was up (of which, with her faulty sense of arithmetic, she could not keep track anyway), and her son had not yet arrived, then, with a mother's implicit faith, with unquestioning credulity she believed without any doubt that her son was busy and accepted the new time almost at once. This was the first time that she openly rebelled.

"You tell me the truth," she repeated quietly but obstinately. She put her glasses on and, bending even nearer to Lucy's face, she bore her dark eyes into those of the young woman. "When is he coming home?"

"Well, in his last letter he said that in three months he will definitely have completed his work. One month is now up," said Lucy.

The old woman suddenly fell back on her pillows, and in a thin, old voice, began to cry. The tears dropped one by one from her open eyes, ran down her narrow, bony face—which remained undistorted by their touch—flowed down her chin and under the ribbon of the black velvet bonnet. Lucy had known her mother-in-law for several years, but she had never seen her cry, not even in 'fifty-six when the shells nearly ripped the roof from over their heads and the two women stuck it out alone for days in the vibrating room because Irene was with her husband at the other end of the town and did not dare come home. They had no news of Janos for a week then. The old woman was lying more or less helplessly in bed even then, but not a muscle moved in her set face at the sound of the detonations—true, she was hard of hearing even then—and all day long she kept telling stories to Lucy about her own youth, about Janos, and made Lucy tell of her own childhood which she spent in the country.

"Please don't cry, Mother," said Lucy, whose heart had stopped in terror for a moment. "Two months is not such a very long time."

"I won't live that long," the old woman said between the two rows of tears which were falling from her eyes. "I won't live till then."

"Oh, Mother, what are you talking about?" Lucy cried with all her strength. "Professor Hetenyi said you will live for a hundred years."

"What do you say?"

"Professor Hetenyi said that . . ."

The old woman turned her tear-stained face toward her daughter-in-law. She did not speak, but stared with her guileless, leaden eyes at Lucy's face for a while, then lifted her hand—and let it fall again.

"That's at least four more years," cried Lucy.

The old woman looked at the ceiling again. "Ach—go away," she said. "Maybe Professor Hetenyi is dead as well."

There was complete silence in the room for a second; only the faint sound of whimpering came from the pillows. "All right, Mother," Lucy said with sudden decision, "I'll write to Janos and tell him to come home."

"What do you say?" asked the old woman after a few seconds.

"I'll write to Janos and tell him to come home at once," repeated Lucy in a louder voice.

"When?"

"Today."

Looking at Lucy again, the old woman said, "I don't ask that. I ask when he should come home?"

"At once!" cried Lucy. "You are perfectly right, Mother, he's been knocking around America long enough. I'm sick of it, too."

The old woman's tears ceased, and a deep sigh escaped her. "That he should come home at once?" she asked dubiously, as if not believing her tears.

"I'll tell him," said Lucy, bending over the bed, "that if he hasn't finished that goddam picture by now, he may as well leave it and come home. You're absolutely right, Mother, I'm not prepared to wait any longer, either."

"That he should leave the film?"

"Don't worry, Mother," said Lucy, "if I tell him to come home or else I'll divorce him, he'll be home in no time."

"You divorce him?"

"And how!" Lucy said. "Well, tell me, Mother, is it fair to leave a woman alone for that long? Everyone else would have been unfaithful to him by now except me. Well, what the hell does he take me for?"

"That he should leave the film?" the old woman said again with trembling lips. "Before the first night?"

"Damn the first night!" said Lucy. "He can stuff his first night. I don't want his dollars. He'd better come home before I run out on him."

The old woman sat up. "What you say? That you will run out from Janos?"

"To hell with his dollars," said Lucy. "I know he won't get a cent if he doesn't finish the picture and that he slaved away for nothing for heaven knows how long, but I don't care. I'm not going to stew here on my own any longer."

The old woman sat up quite straight. "If Janos comes home now then he will not get one dollar?"

"Of course not," said Lucy furiously, watching the old woman from the corner of her eye in the meantime. "So what? I don't want his money. You're quite right, Mother, I'll send him a wire today to come home at once."

The old woman was silent. Then, *"Dumme Gans,"* she said suddenly, her forehead flushing with anger. "And then he has made the whole work for nothing?"

"Made it for nothing," said Lucy. "So what? Work isn't the only thing in life. He's got a wife and an adored mother as well."

"Dumme Gans!" the old woman repeated, getting more and more red in the face. "For man the most important thing is his work. You cannot wear on your conscience that Janos did not finish the film. Do you not know that also the Queen of Holland wants to see?"

Lucy shook her head. "I don't care, I'm going to wire him today. You are absolutely right, Mother."

"I'm not right," screeched the old woman angrily. "You being very stupid. Just because I whine a little bit you do not have to already tear your hair. You shall not write him anything at all."

"I'm not writing, I'm sending a telegram," said Lucy. "Who is more important, Mother, you or the Queen of Holland?"

"That I forbid, that you should telegram," said the old woman. Her convulsed face, which the unusual anguish emaciated to even more angular dimensions, loomed for a moment gauntly and threateningly into Lucy's. But the next instant an almost incredible transformation took place; the face softened with affectionate tenderness, and she smiled at Lucy with all the captivating charm and wisdom of her ninety-odd years.

"You listen to me, my child," she said. "I know what is hurting you. I know you feel jealous for Janos. But I know my son, you be quite sure. My son is like I am, he is faithful to the people he loves. It can happen that he will have one or two affairs in America with women—will have one or two affairs in Europe too, because I had an affair once myself—but in the end my son will tell every woman: Out! You do not have to telegram to him. My son will live to the end of his life with you."

Fortunately the bell rang at that moment. Dr. Illes, the new doctor, looked in at the old woman for a quarter of an hour's therapeutic talk. The old woman—by some mysterious process of the mind, she associated the doctor with her son—was very fond of him, and grumbled at him incessantly. The more she grumbled the more she liked him. Her tear-stained worn face lit up instantly as she now saw him at her bedside. He was a doctor, that was reason enough for deference; the old lady acclaimed science with all the superstitious veneration of the nineteenth century; but apart from that, the six-foot-tall, broad-shouldered young man with the intelligent smile appealed to her still sensitive feminine heart through his masculine charm. His balding head reminded her of Janos, his classical literary education—although it could not compete with Professor Hetenyi's—evoked memories of her own girlhood and beaux. They spattered each other with quotations like playful children bathing in a river who splash water into each other's faces.

There was only one thing the old lady could not bear with equanimity, and that was the fact that Dr. Illes occasionally sang in church at the Sunday service at the church of the Inner City, or at Ascension Day at the church of Kristina Square.

"Well, have you been singing in the church again, Doctor?" she asked petulantly, fixing her tired eyes on the doctor's kindly face. "Do you know that I have not been in church for ninety years?"

"Not once?" asked the doctor. "Not even on your wedding day?"

"Not even then," said the old lady. "Only to look at pictures, when my husband took me to honeymoon in France. But I do not understand well the pictures, so I mainly just walked around in the church and I looked at the kneeling people praying and I thought, *Oh, les pauvres!*"

The old housekeeper, who was standing in the doorway behind the doctor, secretly made the sign of the cross.

"I was ten or twelve years old," said the old woman, "and I have a new dress from my mamma because next day we wanted to make excursion to the Wiener Wald. In the evening I prayed for long to God that he should make fine weather so that we can go for excursion and I can wear my new dress. And then next day it rained. Then I was very angry with God."

"And you are still angry, Mother?" asked Lucy.

The old woman slowly turned her eyes toward Lucy. "I said, my child, that I got very angry then. And the next day when my mamma wanted to take me to church in the morning, I said No."

"And you didn't go?"

"My mamma, she was a clever woman," said the old lady. "She waited until next Sunday when there was sunshine and we could go for excursion and I could wear my new dress and was very happy. And then she asked, Are you coming to church again now, Stina?"

The doctor laughed and patted the old woman's hand, which he had been holding all the time. "And Stina was clever, too, was she?"

The old woman smiled. "My mamma asked, 'Stina, are you coming to church now?' And again I said No."

For a few seconds there was silence in the room. The sun was shining from behind the bed, highlighting the gaunt face molded of skin and bone.

"I said to my mamma," the old lady went on, slightly out of breath, "I said, 'No.' I said, 'I will not ask for help from no one in this world whom you cannot trust. I want to walk on my own feet and head.' "

"Well, that you've achieved," said Lucy in an aside.

"What do you say?" the old woman asked.

"I said, Mother," Lucy shouted, "that you've most certainly achieved that."

The old woman shook her head. "I don't know. . . . I don't know if I achieved. Since I become so old—since then I'm so lonely and helpless. But when I die," she said, "then only my son will hold my hand." She looked at Lucy again. "Not even you. . . . I will not want you then either," she said slowly with a long, penetrating stare at the young woman's face. "The beasts, they want to be alone when they die. If my son cannot be with me and cannot hold my hand then, then I want to be on my own."

Her strength exhausted, she suddenly closed her eyes. The various events of the day had evidently tired her; she withdrew her hand from the doctor's and tucked it under the eiderdown next to the still dependable warmth of her own body. Out of tact for her visitors, she did not turn toward the wall, but pretended deafness, or, maybe, by autosuggestion, became temporarily deaf, and did not reply to any more questions. But when Lucy, following the doctor, wanted to tiptoe out of the room, the old lady lifted her head from her lilac pillow and called her daughter-in-law back.

"Lucy," she said, clearly and intelligibly, "you please to stay with me a little longer."

Lucy floundered. The old lady hardly ever used her Christian name. "Of course, Mother," she said. "I'm only seeing Dr. Illes out."

"Thank you," said the old lady, "but afterwards you come back?"

"Of course I'll come back," said Lucy. "We haven't seen much of each other today."

Despite her tiredness the old lady kept Lucy with her for a long time. She shared her lunch with her while she herself only ate a few morsels—Irene too offered some of her own tomato sauce and braised beef. Lucy ate with such a good appetite and obvious enjoyment, transforming the food so quickly into energy, gaiety, and jokes, that the gastric juices of the other two women began to work at a more healthy rate. They too seemed to thrive on what Lucy put away.

"Is it good?" the old woman asked.

"First-class," said Lucy.

"The tomato sauce as well?"

"Absolutely. If only I liked it!"

"You don't like tomato sauce?"

"Loathe it," said Lucy.

When they had finished the black coffee which the old lady had ordered specially for her guest, Lucy had to undo her waist-length red hair and comb it in front of the bed before the large, silver-framed mirror of the dressing table. The sun rose and shone directly into the room now, onto the mirror, the white arm that rose and fell in front of it combing the Titian red hair with a large ivory comb, coaxing electric sparks invisibly out of its long glittering strands and scattering them into the room. The old woman looked for a while hungrily at the young, slender neck, the white shoulders gleaming in the sun, drank in the healthy fragrance of the mass of lustrous hair—then suddenly she had had enough, and became weary.

"I thank you, now you go home," she said, almost as soon as Lucy had put the last pin into her hair.

The next day the young woman arrived an hour earlier than usual. She rang and started hammering at the door with her two fists. "Irene," she shouted, still outside the door, "Irene! I've got a letter, a letter, a letter . . ." She fell around the housekeeper's neck and, laughing loudly, happily, she turned her around and waltzed along the narrow little hall. She was so excited that her teeth chattered audibly while she danced.

"From your husband?" Irene asked.

"He's alive!" the young woman panted. "It's quite sure now that he is alive. I can go and visit him next Sunday. It was not a letter that came only, but a printed permission form saying that I can go and visit him next Sunday. If he is permitted visitors, then he must be in good health."

Irene did not answer, but the young woman was so happy that she did not notice.

"Even if he isn't in good health, at least he's alive. That much is certain."

"The old lady is very unwell," Irene said after a time.

Lucy sat down on the kitchen stool. Now she noticed the traces of a sleepless night on the old housekeeper's apple cheeks, the shadows under the eyes, the puffed lids. "What happened?" she asked.

"Well, in the middle of the night I woke up—you know how lightly I sleep—because I heard talking in her room. So I went in and she was talking to herself, but much louder than she usually does; she was even shouting and quarreling with her son. At dawn my husband went to phone the doctor, because by then her cheeks were bright red."

"Pneumonia?"

After the onset of the illness the old lady did not lose consciousness for three more days, but, like a poet, by this time she was only concerned with herself. Out of courtesy, and self-discipline, she made an effort to conceal that she had no interest in anything in the world anymore—the last honor her disciplined pride paid to the world of the living—but her efforts were not very successful. She could not hide her indifference very well; the people around her found out soon enough the polite trick she was trying to play.

The inflammation first attacked the left lung, and when they succeeded in subduing it there, the right one; then it spread to both lungs again. While she was conscious the old lady—when she believed herself to be alone—sometimes kept up long monologues, mainly about the small inconveniences of her helpless state. As at the onset of pneumonia the bladder stopped working, she felt very ashamed about having a nurse come twice a day to draw off her water. Later she became unconscious, and did not speak at all. Lucy wanted to hire a full-time nurse, but Irene was having none of it. "What for—so I can look after her, too?" she asked. "Anyway, how do I know what she is going to take out of the cupboard when she is alone in the room?" Apart from this, while she was able to speak or thought speaking worthwhile, the old lady herself protested against the constant presence of a stranger. The first time she spoke was on the afternoon of the third day, during the doctor's visit.

She looked at the visitors, Lucy and the doctor, with clear open eyes, but did not return their greeting, and when they sat down by her bed, she kept her eyes fixed on the door as if expecting someone else who was certain to come. Even later, in her unconscious state she kept this posture; when they straightened her pillows and placed her narrow, black-bonneted head on the topmost one, the next instant she turned it toward the door, craning her neck, and whether her eyes were open or closed, her whole blind body was listening to a sound from that direction. In the last days she even brought her right hand from under the warmth of her eiderdown, and, twisting her wrist at a queer angle, placed her palm under her chin. If Lucy cautiously eased it out of its contorted position and put it back under the eiderdown, the old lady, stubborn even in unconsciousness, slid it out the next moment and wedged it under her chin again. Otherwise she lay as quiet and motionless as a dry leaf in the sunny dust, her breath could neither be seen nor heard, her legs were stretched out one next to the other, the one in plaster touching the other waxen foot. She did not complain even during her unconsciousness; she neither sighed nor groaned; the only time she uttered a little plaintive whimper, so unbefitting to her severe and austere life, was when she was turned on her side so that the doctor, who now came twice a day, could examine her lungs, or the little sores on her back, caused by the long stay in bed, could be dusted with talcum powder. She submitted to the injections without a sound; the hypodermic needle touched only the topmost layer of her mortal form.

But on this last afternoon of her conscious life, she did not reply anymore to the doctor's questions or to Lucy's endearments; compressing her lips and fixing her, for the moment, still clear eyes on the door, she remained obstinately silent.

"I might be able to keep her alive for two or three more days," the doctor said to Lucy after the examination. "Is there any hope . . . ?"

"None."

"What is your wish?" asked the doctor, looking out of the window to the sunflecked lawn and beyond it to the dusty road.

"Whatever is better for her."

"It's all the same for her."

"Whatever is better," said Lucy.

"She does not feel anything by now," the doctor said. "Within a few hours she'll probably lose consciousness. With caffeine, and strofaine, I can keep her heart going for a few more days. Is there any hope for . . . ?"

"No," said Lucy.

"What is your wish?"

"Whatever is best for her."

After the injection, Lucy went out with the doctor into the hall. They had hardly reached it when the bell rang. This was the first time the old lady had used the bell since she had had pneumonia.

"Mother, Mother darling," Lucy whispered, bending over the bed. The doctor was waiting in the hall. The old lady went on looking at the door but clung to Lucy's arm with her hand.

"I did not tell you, my child," she said in a queer muffled voice, "about the last letter of my son. Do you know he is going to get the Kossuth Prize?"

"He didn't write to me about it yet," said Lucy.

The old woman was keeping her eyes on the door. "He is going to get it," she said. "Now you get out and send me in the doctor."

The doctor came in and bent over the bed. "Do you know, singing doctor," said the old lady, still keeping her eyes on the door, but with her mouth twisting into an ironic grimace, "do you know, singing doctor, that my son is going to get the Kossuth Prize?"

"Wonderful," said the doctor, "wonderful."

The old woman nodded again. "Is going to get," she said. "It is a secret not, but you tell Lucy, my daughter-in-law, that she should be proud of my son."

"Wonderful," said the doctor again.

"Now you go on singing," said the old lady.

After the doctor had left, Lucy sat down in the kitchen with Irene.

Her lips were trembling so badly that it was a long time before she could speak. Irene poured her a glass of brandy, put it in front of her and left the bottle out. Lucy put her shaking fingers to the stem of the glass. "Now what am I going to tell him on next Sunday's visit?" she asked.

The Unattainable Other

"*The Birthmark*" suggests another way of organizing a collection like this: taking those imaginary projections of women which have so often passed for the real thing—The Bitch, The Whore (with or without A Heart of Gold), The Angel, and so on—and playing them off against a particular incarnation in fiction or poetry which moves away from the stereotype because all the weight of observation is against a mental construct that nevertheless hovers, a shadow, in the background.

The Hawthorne story belongs to the category of The Unattainable, the woman worshiped precisely because her remoteness from common humanity allows her lover to believe in a permanence that can withstand change and death (into this category also would go Dante's Beatrice and scores of Romantic heroines). Because neither husband nor wife in "The Birthmark" is much of an individual character, the essential dramatic situation stands out boldly. Aylmer is a man whose wife conforms to his ideal vision—with one small exception. He says he loves her but still cannot accept her lapse from that perfection which he conceives in his mind and yearns to project on her living flesh. Georgiana —displaying that classic passivity so extolled as the very definition of femininity by a multitude of writers, philosophers, and psychiatrists— allows him to perform experiments designed to realize this ideal. Although in the beginning she does not wish to have her birthmark removed, she comes to share his vision of it—"Danger? There is but one danger—that this horrible stigma shall be left upon my cheek!"— and goes willingly to the death that must follow upon removal of that last blemish.

There is much more to this story, of course. But how much it is, *above all, a paradigm of the traditional sexual relation, stripped of all the social, domestic and economic outriders that have encumbered it!*

NATHANIEL HAWTHORNE
(1804–1864)

The Birthmark

In the latter part of the last century there lived a man of science, an eminent proficient in every branch of natural philosophy, who not long before our story opens had made experience of a spiritual affinity more attractive than any chemical one. He had left his laboratory to the care of an assistant, cleared his fine countenance from the furnace smoke, washed the stain of acids from his fingers, and persuaded a beautiful woman to become his wife. In those days when the comparatively recent discovery of electricity and other kindred mysteries of Nature seemed to open paths into the region of miracle, it was not unusual for the love of science to rival the love of woman in its depth and absorbing energy. The higher intellect, the imagination, the spirit, and even the heart might all find their congenial aliment in pursuits which, as some of their ardent votaries believed, would ascend from one step of powerful intelligence to another, until the philosopher should lay his hand on the secret of creative force and perhaps make new worlds for himself. We know not whether Aylmer possessed this degree of faith in man's ultimate control over Nature. He had devoted himself, however, too unreservedly to scientific studies ever to be weaned from them by any second passion. His love for his young wife might prove the stronger of the two; but it could only be by intertwining itself with his love of science, and uniting the strength of the latter to his own.

Such a union accordingly took place, and was attended with truly remarkable consequences and a deeply impressive moral. One day, very soon after their marriage, Aylmer sat gazing at his wife with a trouble in his countenance that grew stronger until he spoke.

"Georgiana," said he, "has it never occurred to you that the mark upon your cheek might be removed?"

"No, indeed," said she, smiling; but perceiving the seriousness of his manner, she blushed deeply. "To tell you the truth it has been so often called a charm that I was simple enough to imagine it might be so."

"Ah, upon another face perhaps it might," replied her husband; "but never on yours. No, dearest Georgiana, you came so nearly perfect from the hand of Nature that this slightest possible defect, which we hesitate whether to term a defect or a beauty, shocks me, as being the visible mark of earthly imperfection."

"Shocks you, my husband!" cried Georgiana, deeply hurt; at first reddening with momentary anger, but then bursting into tears. "Then why did you take me from my mother's side? You cannot love what shocks you!"

To explain this conversation it must be mentioned that in the centre of Georgiana's left cheek there was a singular mark, deeply inter- woven, as it were, with the texture and substance of her face. In the usual state of her complexion—a healthy though delicate bloom—the mark wore a tint of deeper crimson, which imperfectly defined its shape amid the surrounding rosiness. When she blushed it gradually became more indistinct, and finally vanished amid the triumphant rush of blood that bathed the whole cheek with its brilliant glow. But if any shifting motion caused her to turn pale there was the mark again, a crimson stain upon the snow, in what Aylmer sometimes deemed an almost fearful distinctness. Its shape bore not a little simi- larity to the human hand, though of the smallest pygmy size. Georgi- ana's lovers were wont to say that some fairy at her birth hour had laid her tiny hand upon the infant's cheek, and left this impress there in token of the magic endowments that were to give her such sway over all hearts. Many a desperate swain would have risked life for the privilege of pressing his lips to the mysterious hand. It must not be concealed, however, that the impression wrought by this fairy sign manual varied exceedingly, according to the difference of tempera- ment in the beholders. Some fastidious persons—but they were exclusively of her own sex—affirmed that the bloody hand, as they chose to call it, quite destroyed the effect of Georgiana's beauty, and rendered her countenance even hideous. But it would be as reasonable to say that one of those small blue stains which sometimes occur in the purest statuary marble would convert the Eve of Powers to a monster. Masculine observers, if the birthmark did not heighten their admiration, contented themselves with wishing it away, that the world might possess one living specimen of ideal loveliness without the semblance of a flaw. After his marriage—for he thought little or nothing of the matter before—Aylmer discovered that this was the case with himself.

Had she been less beautiful—if Envy's self could have found aught else to sneer at—he might have felt his affection heightened by the prettiness of this mimic hand, now vaguely portrayed, now lost, now stealing forth again and glimmering to and fro with every pulse of emotion that throbbed within her heart; but seeing her otherwise so perfect, he found this one defect grow more and more intolerable with every moment of their united lives. It was the fatal flaw of humanity which Nature, in one shape or another, stamps ineffaceably on all her productions, either to imply that they are temporary and finite, or that their perfection must be wrought by toil and pain. The crimson hand expressed the ineludible gripe in which mortality clutches the highest and purest of earthly mould, degrading them into kindred with the lowest, and even with the very brutes, like whom their visible frames return to dust. In this manner, selecting it as the symbol of his wife's liability to sin, sorrow, decay, and death, Aylmer's sombre imagination was not long in rendering the birthmark a frightful object, causing him more trouble and horror than ever Georgiana's beauty, whether of soul or sense, had given him delight.

At all the seasons which should have been their happiest, he invariably and without intending it, nay, in spite of a purpose to the contrary, reverted to this one disastrous topic. Trifling as it at first appeared, it so connected itself with innumerable trains of thought and modes of feeling that it became the central point of all. With the morning twilight Aylmer opened his eyes upon his wife's face and recognized the symbol of imperfection; and when they sat together at the evening hearth his eyes wandered stealthily to her cheek, and beheld, flickering with the blaze of the wood fire, the spectral hand that wrote mortality where he would fain have worshipped. Georgiana soon learned to shudder at his gaze. It needed but a glance with the peculiar expression that his face often wore to change the roses of her cheek into a deathlike paleness, amid which the crimson hand was brought strongly out, like a bas-relief of ruby on the whitest marble.

Late one night when the lights were growing dim, so as hardly to betray the stain on the poor wife's cheek, she herself, for the first time, voluntarily took up the subject.

"Do you remember, my dear Aylmer," said she, with a feeble attempt at a smile, "have you any recollection of a dream last night about this odious hand?"

"None! none whatever!" replied Aylmer, starting; but then he added, in a dry, cold tone, affected for the sake of concealing the real depth of his emotion, "I might well dream of it; for before I fell asleep it had taken a pretty firm hold of my fancy."

"And you did dream of it!" continued Georgiana, hastily; for she dreaded lest a gush of tears should interrupt what she had to say. "A terrible dream! I wonder that you can forget it. Is it possible to forget this one expression?—'It is in her heart now; we must have it out!' Reflect, my husband; for by all means I would have you recall that dream."

The mind is in a sad state when Sleep, the all-involving, cannot confine her spectres within the dim region of her sway, but suffers them to break forth, affrighting this actual life with secrets that perchance belong to a deeper one. Aylmer now remembered his dream. He had fancied himself with his servant Aminadab, attempting an operation for the removal of the birthmark; but the deeper went the knife, the deeper sank the hand, until at length its tiny grasp appeared to have caught hold of Georgiana's heart; whence, however, her husband was inexorably resolved to cut or wrench it away.

When the dream had shaped itself perfectly in his memory, Aylmer sat in his wife's presence with a guilty feeling. Truth often finds its way to the mind close muffled in robes of sleep, and then speaks with uncompromising directness of matters in regard to which we practise an unconscious self-deception during our waking moments. Until now he had not been aware of the tyrannizing influence acquired by one idea over his mind, and of the lengths which he might find in his heart to go for the sake of giving himself peace.

"Aylmer," resumed Georgiana, solemnly, "I know not what may be the cost to both of us to rid me of this fatal birthmark. Perhaps its removal may cause cureless deformity; or it may be the stain goes as deep as life itself. Again: do we know that there is a possibility, on any terms, of unclasping the firm gripe of this little hand which was laid upon me before I came into the world?"

"Dearest Georgiana, I have spent much thought upon the subject," hastily interrupted Aylmer. "I am convinced of the perfect practicability of its removal."

"If there be the remotest possibility of it," continued Georgiana, "let the attempt be made at whatever risk. Danger is nothing to me; for life, while this hateful mark makes me the object of your horror and disgust—life is a burden which I would fling down with joy. Either remove this dreadful hand, or take my wretched life! You have deep science. All the world bears witness of it. You have achieved great wonders. Cannot you remove this little, little mark, which I cover with the tips of two small fingers? Is this beyond your power, for the sake of your own peace, and to save your poor wife from madness?"

"Noblest, dearest, tenderest wife," cried Aylmer, rapturously, "doubt not my power. I have already given this matter the deepest thought—thought which might almost have enlightened me to create a being less perfect than yourself. Georgiana, you have led me deeper than ever into the heart of science. I feel myself fully competent to render this dear cheek as faultless as its fellow; and then, most beloved, what will be my triumph when I shall have corrected what Nature left imperfect in her fairest work! Even Pygmalion, when his sculptured woman assumed life, felt not greater ecstasy than mine will be."

"It is resolved, then," said Georgiana, faintly smiling. "And, Aylmer, spare me not, though you should find the birthmark take refuge in my heart at last."

Her husband tenderly kissed her cheek—her right cheek—not that which bore the impress of the crimson hand.

The next day Aylmer apprised his wife of a plan that he had formed whereby he might have opportunity for the intense thought and constant watchfulness which the proposed operation would require; while Georgiana, likewise, would enjoy the perfect repose essential to its success. They were to seclude themselves in the extensive apartments occupied by Aylmer as a laboratory, and where, during his toilsome youth, he had made discoveries in the elemental powers of Nature that had roused the admiration of all the learned societies in Europe. Seated calmly in this laboratory, the pale philosopher had investigated the secrets of the highest cloud region and of the profoundest mines; he had satisfied himself of the causes that kindled and kept alive the fires of the volcano; and had explained the mystery of fountains, and how it is that they gush forth, some so bright and pure, and others with such rich medicinal virtues, from the dark bosom of the earth. Here, too, at an earlier period, he had studied the wonders of the human frame, and attempted to fathom the very process by which Nature assimilates all her precious influences from earth and air, and from the spiritual world, to create and foster man, her masterpiece. The latter pursuit, however, Aylmer had long laid aside in unwilling recognition of the truth—against which all seekers sooner or later stumble—that our great creative Mother, while she amuses us with apparently working in the broadest sunshine, is yet severely careful to keep her own secrets, and, in spite of her pretended openness, shows us nothing but results. She permits us, indeed, to mar, but seldom to mend, and, like a jealous patentee, on no account to make. Now, however, Aylmer resumed these half-forgotten investigations; not, of course, with such hopes or wishes as first suggested them; but because they involved much physiological truth and lay in the path of his proposed scheme for the treatment of Georgiana.

As he led her over the threshold of the laboratory, Georgiana was cold and tremulous. Aylmer looked cheerfully into her face, with intent to reassure her, but was so startled with the intense glow of the birthmark upon the whiteness of her cheek that he could not restrain a strong convulsive shudder. His wife fainted.

"Aminadab! Aminadab!" shouted Aylmer, stamping violently on the floor.

Forthwith there issued from an inner apartment a man of low stature, but bulky frame, with shaggy hair hanging about his visage, which was grimed with the vapors of the furnace. This personage had been Aylmer's underworker during his whole scientific career, and was admirably fitted for that office by his great mechanical readiness, and the skill with which, while incapable of comprehending a single principle, he executed all the details of his master's experiments. With his vast strength, his shaggy hair, his smoky aspect, and the indescribable earthiness that incrusted him, he seemed to represent man's physical nature; while Aylmer's slender figure, and pale, intellectual face, were no less apt a type of the spiritual element.

"Throw open the door of the boudoir, Aminadab," said Aylmer, "and burn a pastil."

"Yes, master," answered Aminadab, looking intently at the lifeless form of Georgiana; and then he muttered to himself, "If she were my wife, I'd never part with that birthmark."

When Georgiana recovered consciousness she found herself breathing an atmosphere of penetrating fragrance, the gentle potency of which had recalled her from her deathlike faintness. The scene around her looked like enchantment. Aylmer had converted those smoky, dingy, sombre rooms, where he had spent his brightest years in recondite pursuits, into a series of beautiful apartments not unfit to be the secluded abode of a lovely woman. The walls were hung with gorgeous curtains, which imparted the combination of grandeur and grace that no other species of adornment can achieve; and as they fell from the ceiling to the floor, their rich and ponderous folds, concealing all angles and straight lines, appeared to shut in the scene from infinite space. For aught Georgiana knew, it might be a pavilion among the clouds. And Aylmer, excluding the sunshine, which would have interfered with his chemical processes, had supplied its place with perfumed lamps, emitting flames of various hue, but all uniting in a soft, impurpled radiance. He now knelt by his wife's side, watching her earnestly, but without alarm; for he was confident in his science, and felt that he could draw a magic circle round her within which no evil might intrude.

"Where am I? Ah, I remember," said Georgiana, faintly; and she placed her hand over her cheek to hide the terrible mark from her husband's eyes.

"Fear not, dearest!" exclaimed he. "Do not shrink from me! Believe me, Georgiana, I even rejoice in this single imperfection, since it will be such a rapture to remove it."

"Oh, spare me!" sadly replied his wife. "Pray do not look at it again. I never can forget that convulsive shudder."

In order to soothe Georgiana, and, as it were, to release her mind from the burden of actual things, Aylmer now put in practise some of the light and playful secrets which science had taught him among its profounder lore. Airy figures, absolutely bodiless ideas, and forms of unsubstantial beauty came and danced before her, imprinting their momentary footsteps on beams of light. Though she had some indistinct idea of the method of these optical phenomena, still the illusion was almost perfect enough to warrant the belief that her husband possessed sway over the spiritual world. Then again, when she felt a wish to look forth from her seclusion, immediately, as if her thoughts were answered, the procession of external existence flitted across a screen. The scenery and the figures of actual life were perfectly represented, but with that bewitching, yet indescribable difference which always makes a picture, an image, or a shadow so much more attractive than the original. When wearied of this, Aylmer bade her cast her eyes upon a vessel containing a quantity of earth. She did so, with little interest at first; but was soon startled to perceive the germ of a plant shooting upward from the soil. Then came the slender stalk; the leaves gradually unfolded themselves; and amid them was a perfect and lovely flower.

"It is magical!" cried Georgiana. "I dare not touch it."

"Nay, pluck it," answered Aylmer—"pluck it, and inhale its brief perfume while you may. The flower will wither in a few moments and leave nothing save its brown seed vessels; but thence may be perpetuated a race as ephemeral as itself."

But Georgiana had no sooner touched the flower than the whole plant suffered a blight, its leaves turning coal-black as if by the agency of fire.

"There was too powerful a stimulus," said Aylmer, thoughtfully.

To make up for this abortive experiment, he proposed to take her portrait by a scientific process of his own invention. It was to be effected by rays of light striking upon a polished plate of metal. Georgiana assented; but, on looking at the result, was affrighted to find the features of the portrait blurred and indefinable; while the minute figure of a hand appeared where the cheek should have been. Aylmer snatched the metallic plate and threw it into a jar of corrosive acid.

Soon, however, he forgot these mortifying failures. In the intervals of study and chemical experiment he came to her flushed and exhausted, but seemed invigorated by her presence, and spoke in glowing language of the resources of his art. He gave a history of the long dynasty of the alchemists, who spent so many ages in quest of the universal solvent by which the golden principle might be elicited from all things vile and base. Aylmer appeared to believe that, by the plainest scientific logic, it was altogether within the limits of possibility to discover this long-sought medium; "but," he added, "a philosopher who should go deep enough to acquire the power would attain too lofty a wisdom to stoop to the exercise of it." Not less singular were his opinions in regard to the elixir vitæ. He more than intimated that it was at his option to concoct a liquid that should prolong life for years, perhaps interminably; but that it would produce a discord in Nature which all the world, and chiefly the quaffer of the immortal nostrum, would find cause to curse.

"Aylmer, are you in earnest?" asked Georgiana, looking at him with amazement and fear. "It is terrible to possess such power, or even to dream of possessing it."

"Oh, do not tremble, my love," said her husband. "I would not wrong either you or myself by working such inharmonious effects upon our lives; but I would have you consider how trifling, in comparison, is the skill requisite to remove this little hand."

At the mention of the birthmark, Georgiana, as usual, shrank as if a redhot iron had touched her cheek.

Again Aylmer applied himself to his labors. She could hear his voice in the distant furnace room giving directions to Aminadab, whose harsh, uncouth, misshapen tones were audible in response, more like the grunt or growl of a brute than human speech. After hours of absence, Aylmer reappeared and proposed that she should now examine his cabinet of chemical products and natural treasures of the earth. Among the former he showed her a small vial, in which, he remarked, was contained a gentle yet most powerful fragrance, capable of impregnating all the breezes that blow across a kingdom. They were of inestimable value, the contents of that little vial; and, as he said so, he threw some of the perfume into the air and filled the room with piercing and invigorating delight.

"And what is this?" asked Georgiana, pointing to a small crystal globe containing a gold-colored liquid. "It is so beautiful to the eye that I could imagine it the elixir of life."

"In one sense it is," replied Aylmer; "or, rather, the elixir of immortality. It is the most precious poison that ever was concocted in this world. By its aid I could apportion the lifetime of any mortal at whom you might point your finger. The strength of the dose would determine whether he were to linger out years, or drop dead in the midst of a breath. No king on his guarded throne could keep his life if I, in my private station, should deem that the welfare of millions justified me in depriving him of it."

"Why do you keep such a terrific drug?" inquired Georgiana in horror.

"Do not mistrust me, dearest," said her husband, smiling; "its virtuous potency is yet greater than its harmful one. But see! here is a powerful cosmetic. With a few drops of this in a vase of water, freckles may be washed away as easily as the hands are cleansed. A stronger infusion would take the blood out of the cheek, and leave the rosiest beauty a pale ghost."

"Is it with this lotion that you intend to bathe my cheek?" asked Georgiana, anxiously.

"Oh, no," hastily replied her husband; "this is merely superficial. Your case demands a remedy that shall go deeper."

In his interviews with Georgiana, Aylmer generally made minute inquiries as to her sensations and whether the confinement of the rooms and the temperature of the atmosphere agreed with her. These questions had such a particular drift that Georgiana began to conjecture that she was already subjected to certain physical influences, either breathed in with the fragrant air or taken with her food. She fancied likewise, but it might be altogether fancy, that there was a stirring up of her system—a strange, indefinite sensation creeping through her veins, and tingling, half painfully, half pleasurably, at her heart. Still, whenever she dared to look into the mirror, there she beheld herself pale as a white rose and with the crimson birthmark stamped upon her cheek. Not even Aylmer now hated it so much as she.

To dispel the tedium of the hours which her husband found it necessary to devote to the processes of combination and analysis, Georgiana turned over the volumes of his scientific library. In many dark old tomes she met with chapters full of romance and poetry. They were the works of philosophers of the middle ages, such as Albertus Magnus, Cornelius Agrippa, Paracelsus, and the famous friar who created the prophetic Brazen Head. All these antique naturalists stood in advance of their centuries, yet were imbued with some of their credulity, and therefore were believed, and perhaps imagined themselves to have acquired from the investigation of Nature a power about Nature, and from physics a sway over the spiritual world. Hardly less curious and imaginative were the early volumes of the Transactions of the Royal Society, in which the members, knowing little of the limits of natural possibility, were continually recording wonders or proposing methods whereby wonders might be wrought.

But to Georgiana the most engrossing volume was a large folio from her husband's own hand, in which he had recorded every experiment of his scientific career, its original aim, the methods adopted for its development, and its final success or failure, with the circumstances to which either event was attributable. The book, in truth, was both the history and emblem of his ardent, ambitious, imaginative, yet practical and laborious life. He handled physical details as if there were nothing beyond them; yet spiritualized them all, and redeemed himself from materialism by his strong and eager aspiration towards the infinite. In his grasp the veriest clod of earth assumed a soul. Georgiana, as she read, reverenced Aylmer and loved him more profoundly than ever, but with a less entire dependence on his judgment than heretofore. Much as he had accomplished, she could not but observe that his most splendid successes were almost invariably failures, if compared with the ideal at which he aimed. His brightest diamonds were the merest pebbles, and felt to be so by himself, in comparison with the inestimable gems which lay hidden beyond his reach. The volume, rich with achievements that had won renown for its author, was yet as melancholy a record as ever mortal hand had penned. It was the sad confession and continual exemplification of the shortcomings of the composite man, the spirit burdened with clay and working in matter, and of the despair that assails the higher nature at finding itself so miserably thwarted by the earthly part. Perhaps every man of genius in whatever sphere might recognize the image of his own experience in Aylmer's journal.

So deeply did these reflections affect Georgiana that she laid her face upon the open volume and burst into tears. In this situation she was found by her husband.

"It is dangerous to read in a sorcerer's books," said he with a smile, though his countenance was uneasy and displeased. "Georgiana, there are pages in that volume which I can scarcely glance over and keep my senses. Take heed lest it prove as detrimental to you."

"It has made me worship you more than ever," said she.

"Ah, wait for this one success," rejoined he, "then worship me if you will. I shall deem myself hardly unworthy of it. But come, I have sought you for the luxury of your voice. Sing to me, dearest."

So she poured out the liquid music of her voice to quench the thirst of his spirit. He then took his leave with a boyish exuberance of gayety, assuring her that her seclusion would endure but a little longer, and that the result was already certain. Scarcely had he departed when Georgiana felt irresistibly impelled to follow him. She had forgotten to inform Aylmer of a symptom which for two or three hours past had begun to excite her attention. It was a sensation in the fatal birthmark, not painful, but which induced a restlessness throughout her system. Hastening after her husband, she intruded for the first time into the laboratory.

The first thing that struck her eye was the furnace, that hot and feverish worker, with the intense glow of its fire, which by the quantities of soot clustered above it seemed to have been burning for ages. There was a distilling apparatus in full operation. Around the moon were retorts, tubes, cylinders, crucibles, and other apparatus of chemical research. An electrical machine stood ready for immediate use. The atmosphere felt oppressively close, and was tainted with gaseous odors which had been tormented forth by the processes of science. The severe and homely simplicity of the apartment, with its naked walls and brick pavement, looked strange, accustomed as Georgiana had become to the fantastic elegance of her boudoir. But what chiefly, indeed almost solely, drew her attention, was the aspect of Aylmer himself.

He was pale as death, anxious and absorbed, and hung over the furnace as if it depended upon his utmost watchfulness whether the liquid which it was distilling should be the draught of immortal happiness or misery. How different from the sanguine and joyous mien that he had assumed for Georgiana's encouragement!

"Carefully now, Aminadab; carefully, thou human machine; carefully, thou man of clay!" muttered Aylmer, more to himself than his assistant. "Now, if there be a thought too much or too little, it is all over."

"Ho! ho!" mumbled Aminadab. "Look, master! look!"

Aylmer raised his eyes hastily, and at first reddened, then grew paler than ever, on beholding Georgiana. He rushed towards her and seized her arm with a gripe that left the print of his fingers upon it.

"Why do you come hither? Have you no trust in your husband?" cried he, impetuously. "Would you throw the blight of that fatal birthmark over my labors? It is not well done. Go, prying woman, go!"

"Nay, Aylmer," said Georgiana with the firmness of which she possessed no stinted endowment, "it is not you that have a right to complain. You mistrust your wife; you have concealed the anxiety with which you watch the development of this experiment. Think not so unworthily of me, my husband. Tell me all the risk we run, and fear not that I shall shrink; for my share in it is far less than your own."

"No, no, Georgiana!" said Aylmer, impatiently; "it must not be."

"I submit," replied she calmly. "And, Aylmer, I shall quaff whatever draught you bring me; but it will be on the same principle that would induce me to take a dose of poison if offered by your hand."

"My noble wife," said Aylmer, deeply moved. "I knew not the height and depth of your nature until now. Nothing shall be concealed. Know, then, that this crimson hand, superficial as it seems, has clutched its grasp into your being with a strength of which I had no previous conception. I have already administered agents powerful enough to do aught except to change your entire physical system. Only one thing remains to be tried. If that fail us we are ruined."

"Why did you hesitate to tell me this?" asked she.

"Because, Georgiana," said Aylmer, in a low voice, "there is danger."

"Danger? There is but one danger—that this horrible stigma shall be left upon my cheek!" cried Georgiana. "Remove it, remove it, whatever be the cost, or we shall both go mad!"

"Heaven knows your words are too true," said Aylmer, sadly. "And now, dearest, return to your boudoir. In a little while all will be tested."

He conducted her back and took leave of her with a solemn tenderness which spoke far more than his words how much was now at stake. After his departure Georgiana became rapt in musings. She considered the character of Aylmer, and did it completer justice than at any previous moment. Her heart exulted, while it trembled, at his honorable love—so pure and lofty that it would accept nothing less than perfection nor miserably make itself contented with an earthlier nature than he had dreamed of. She felt how much more precious was such a sentiment than the meaner kind which would have borne with the imperfection for her sake, and have been guilty of treason to holy love by degrading its perfect idea to the level of the actual; and with her whole spirit she prayed that, for a single moment, she might satisfy his highest and deepest conception. Longer than one moment she well knew it could not be; for his spirit was ever on the march, ever ascending, and each instant required something that was beyond the scope of the instant before.

The sound of her husband's footsteps aroused her. He bore a crystal goblet containing a liquor colorless as water, but bright enough to be the draught of immortality. Aylmer was pale; but it seemed rather the consequence of a highly wrought state of mind and tension of spirit than of fear or doubt.

"The concoction of the draught has been perfect," said he, in answer to Georgiana's look. "Unless all my science have deceived me, it cannot fail."

"Save on your account, my dearest Aylmer," observed his wife, "I might wish to put off this birthmark of mortality by relinquishing mortality itself in preference to any other mode. Life is but a sad possession to those who have attained precisely the degree of moral advancement at which I stand. Were I weaker and blinder it might be happiness. Were I stronger, it might be endured hopefully. But, being what I find myself, methinks I am of all mortals the most fit to die."

"You are fit for heaven without tasting death!" replied her husband. "But why do we speak of dying? The draught cannot fail. Behold its effect upon this plant."

On the window seat there stood a geranium diseased with yellow blotches, which had overspread all its leaves. Aylmer poured a small quantity of the liquid upon the soil in which it grew. In a little time, when the roots of the plant had taken up the moisture, the unsightly blotches began to be extinguished in a living verdure.

"There needed no proof," said Georgiana, quietly. "Give me the goblet. I joyfully stake all upon your word."

"Drink, then, thou lofty creature!" exclaimed Aylmer, with fervid admiration. "There is no taint of imperfection on thy spirit. Thy sensible frame, too, shall soon be all perfect."

She quaffed the liquid and returned the goblet to his hand.

"It is grateful," said she with a placid smile. "Methinks it is like water from a heavenly fountain; for it contains I know not what of unobtrusive fragrance and deliciousness. It allays a feverish thirst that had parched me for many days. Now, dearest, let me sleep. My earthly senses are closing over my spirit like the leaves around the heart of a rose at sunset."

She spoke the last words with a gentle reluctance, as if it required almost more energy than she could command to pronounce the faint and lingering syllables. Scarcely had they loitered through her lips ere she was lost in slumber. Aylmer sat by her side, watching her aspect with the emotions proper to a man the whole value of whose existence was involved in the process now to be tested. Mingled with this mood, however, was the philosophic investigation characteristic of the man of science. Not the minutest symptom escaped him. A heightened flush of the cheek, a slight irregularity of breath, a quiver of the eyelid, a hardly perceptible tremor through the frame—such were the details which, as the moments passed, he wrote down in his folio volume. Intense thought had set its stamp upon every previous page of that volume, but the thoughts of years were all concentrated upon the last.

While thus employed, he failed not to gaze often at the fatal hand, and not without a shudder. Yet once, by a strange and unaccountable impulse, he pressed it with his lips. His spirit recoiled, however, in the very act; and Georgiana, out of the midst of her deep sleep, moved uneasily and murmured as if in remonstrance. Again Aylmer resumed his watch. Nor was it without avail. The crimson hand, which at first had been strongly visible upon the marble paleness of Georgiana's cheek, now grew more faintly outlined. She remained not less pale than ever; but the birthmark, with every breath that came and went, lost somewhat of its former distinctness. Its presence had been awful; its departure was more awful still. Watch the stain of the rainbow fading out the sky, and you will know how that mysterious symbol passed away.

"By Heaven! it is well-nigh gone!" said Aylmer to himself, in almost irrepressible ecstasy. "I can scarcely trace it now. Success! success! And now it is like the faintest rose color. The lightest flush of blood across her cheek would overcome it. But she is so pale!"

He drew aside the window curtain and suffered the light of natural day to fall into the room and rest upon her cheek. At the same time he heard a gross, hoarse chuckle, which he had long known as his servant Aminadab's expression of delight.

"Ah, clod; ah, earthly mass!" cried Aylmer, laughing in a sort of frenzy, "you have served me well! Matter and spirit—earth and heaven

—have both done their part in this! Laugh, thing of the senses! You have earned the right to laugh."

These exclamations broke Georgiana's sleep. She slowly unclosed her eyes and gazed into the mirror which her husband had arranged for that purpose. A faint smile flitted over her lips when she recognized how barely perceptible was now that crimson hand which had once blazed forth with such disastrous brilliancy as to scare away all their happiness. But then her eyes sought Aylmer's face with a trouble and anxiety that he could by no means account for.

"My poor Aylmer!" murmured she.

"Poor? Nay, richest, happiest, most favored!" exclaimed he. "My peerless bride, it is successful! You are perfect!"

"My poor Aylmer," she repeated, with a more than human tenderness, "you have aimed loftily; you have done nobly. Do not repent that with so high and pure a feeling, you have rejected the best the earth could offer. Aylmer, dearest Aylmer, I am dying!"

Alas! it was too true! The fatal hand had grappled with the mystery of life, and was the bond by which an angelic spirit kept itself in union with a mortal frame. As the last crimson tint of the birthmark—that sole token of human imperfection—faded from her cheek, the parting breath of the now perfect woman passed into the atmosphere, and her soul, lingering a moment near her husband, took its heavenward flight. Then a hoarse, chuckling laugh was heard again! Thus ever does the gross fatality of earth exult in its invariable triumph over the immortal essence which, in this dim sphere of half development, demands the completeness of a higher state. Yet, had Aylmer reached a profounder wisdom, he need not thus have flung away the happiness which would have woven his mortal life of the selfsame texture with the celestial. The momentary circumstance was too strong for him; he failed to look beyond the shadowy scope of time, and, living once for all in eternity, to find the perfect future in the present.

The Woman Who Lives Inside the World

BY MICHELE MURRAY

I

I am the one who is inside you
kerchiefed
uttering a blessing:
may your eyes feast on arenas of sound
may you breathe like clouds
traveling beyond the sun
May your bones make a latticework
in which I can move outward
into your eyes
into your mouth
into the heart of your life.

2

"I accept the universe."—Margaret Fuller

At times I turn my back on such raw flesh
close my eyes on the gobbets of fat
break the blood into parables of silence

The world continues:
                  or does it wait
upon my wink?

The sun enters its sky cupola
remember me.

3

I am growing into your husk
like wheat grains under a careless sun
Your skin will plump & stretch with the filling of me
you will thin & grow transparent
I will see the sky through your blessing of skin
Your sheath will split like the bodies of the drowned

Fragments of the new will flicker upon the air.

4

Break open the loaf of yourself
I am waiting where there is no eating or drinking.

5

What! Run from the bits of flying paper,
the grit? Flee from the embattled pavements
the whittled crowds? Slip into the pine trees
somewhere away? Snakes, my friends, come to greet me
Their soundless music trembles the earth

The insides of mountains are hollow
They wait for me to slip into them
The dark cones scatter their needles
on the cold streets.

6

Where I crouch there is massive quiet
Covered by silence the lithe jaguar tautens:
leaps into the broken air.

7

Something keeps me from screaming
down the long straight roads
unrolling behind me
Your hand thrust between my teeth perhaps
or a memory of kinder times:
rainfalls of sleep
a time of the dropping of masks
The scream floats inside me
I am inside you
The scream rides with us across the corners of the world.

8

I am still making my body in the dark
The vixen loans me her stinking pelt
When I pull it up my head grows a muzzle
my sharp nose inhales the world
At the end of my paws
pointed nails
sever the sky.

9

Maggots dance inside the sheep's head
They sculpt the eye bulging from its socket
The bones stand forth

The rib bones of the world crack open
Springing from the spreading arc
Aphrodite rises
shaking off milk & blood
Her feet trample the maggots
The bones dance.

10

Turn your head
Look down
I am holding a scarlet thread out to you
Come in.

SUGGESTED FURTHER READING

There is far more material than could be included here; I've limited myself, with a few exceptions, to *literature* and its ancillary subjects.

FICTION

Nineteenth Century and Earlier

Alcott, Louisa May: *Little Women, Little Men, Eight Cousins, Rose in Bloom, An Old-fashioned Girl.*

Austen, Jane: *Pride and Prejudice, Sense and Sensibility, Mansfield Park, Northanger Abbey, Persuasion, Emma.*

Brontë, Charlotte: *Villette, Jane Eyre.*

Brontë, Emily: *Wuthering Heights.*

Burney, Fanny: *Evelina.*

Chekhov, Anton: many, many selections and collections of his stories, with their superbly drawn women characters; a good recent book is *The Image of Chekhov,* edited and translated by Robert Payne (New York: Alfred A. Knopf, 1963).

Chopin, Kate: *The Awakening.*

Defoe, Daniel: *Moll Flanders.*

Dickens, Charles: *Great Expectations, Our Mutual Friend, Bleak House, Little Dorrit.*

Eliot, George: *Adam Bede, Middlemarch, The Mill on the Floss.*

Fielding, Henry: *Amelia.*

Flaubert, Gustave: *Madame Bovary, The Sentimental Education,* "A Simple Heart."

Fontane, Theodor: *Beyond Recall, Effie Briest.* German novelist of the late nineteenth century.

Gaskell, Mary: *Cranford, Wives and Daughters.*

Hardy, Thomas: *Jude the Obscure, The Return of the Native, Tess of the d'Urbervilles, Far From the Madding Crowd.*

Hawthorne, Nathaniel: *The Scarlet Letter, The Marble Faun, The Blithedale Romance.*

James, Henry: *Washington Square, Portrait of a Lady, The Bostonians, The Princess Casamassima, Spoils of Poynton, What Maisie Knew, The Wings of the Dove, The Ambassadors, The Golden Bowl.*

Jewett, Sarah Orne: *The Country of the Pointed Firs.*

Laclos, Choderlos de: *Dangerous Acquaintances.*

Lafayette, Mme de: *Princess of Cleves.*

Meredith, George: *Diana of the Crossways, The Romantic Comedians.*

Murasaki, Lady: *Tale of Genji* (translated by Arthur Waley).

Richardson, Samuel: *Clarissa.*

Stendhal (Henri Beyle): *The Red and the Black, The Charterhouse of Parma, Lucien Leuwen.*

Thackeray, William Makepeace: *Vanity Fair.*

Tolstoy, Leo: *Family Happiness, War and Peace, Anna Karenina, The Kreutzer Sonata.*

Wharton, Edith: *Ethan Frome, Hudson River Bracketed, The Age of Innocence.*

Early Twentieth Century

Bennett, Arnold: *Old Wives' Tale.*

Dreiser, Theodore: *Sister Carrie.*

Forster, E. M.: *Howard's End, A Passage to India.*

Gide, Andre: *Strait Is the Gate.*

Glasgow, Ellen: *Barren Ground, In This Our Life, Virginia, The Sheltered Life.*

Joyce, James: *Ulysses.*

Lawrence, D. H.: *Sons and Lovers, The Rainbow, Women in Love, Lady Chatterley's Lover, Complete Short Stories.*

Mansfield, Katherine: *Short Stories.*

Musil, Robert: *Five Women, The Man Without Qualities.*

More or Less Contemporaries

Arnow, Harriette: *The Dollmaker.*

Ashton-Warner, Sylvia: *Spinster.*

Babel, Isaac: *Collected Stories, The Lonely Years, You Must Know Everything.*

Barnes, Djuna: *Nightwood.*

Bedford, Sibyl: *The Legacy.*

Bowen, Elizabeth: *Death of the Heart, The House in Paris, A World of Love, The Heat of the Day, The Little Girls, Eva Trout.*

Boyle, Kay: *Generation Without Farewell, Monday Night, Plagued by Nightingales, Primer for Combat, Year Before Last.*

Cary, Joyce: *Herself Surprised, A Fearful Joy.*

Cather, Willa: *A Lost Lady, Song of the Lark, My Ántonia, O Pioneers!, The Old Beauty and others.*

Colette: Her works have been issued in a number of translations and in various collections, as well as in a uniform edition being published by Farrar, Straus & Giroux. *Earthly Paradise,* edited by Robert Phelps (Farrar) is a good place to begin. Some of her other works, fiction and nonfiction, are: *Claudine in Paris* (and its sequels), *Mitsou, My Mother's House, The Vagabond, The Shackle, The Ripening Seed, Gigi, Julie de Carneilhan, Stories, Break of Day, Blue Lantern, The Cat, Duo, Chéri, The Last of Chéri, The Indulgent Husband, The Pure and the Impure.*

Compton-Burnett, Ivy: *Bullivant and the Lambs, A God and His Gifts, Two Worlds and Their Ways, The Mighty and their Fall,* and many more.

Connell, Evan S., Jr.: *Mrs. Bridge.*

Dinesen, Isak (Karen Blixen): *Seven Gothic Tales, Winter's Tales, Last Tales, Anecdotes of Destiny, Out of Africa.*

Drabble, Margaret: *A Summer Bird-Cage, The Garrick Year, The Millstone, Jerusalem the Golden, The Waterfall, The Needle's Eye.*

Feinstein, Elaine: *The Circle.*

Gordimer, Nadine: *The Lying Days, A World of Strangers, Occasion for Loving, The Late Bourgeois World, A Guest of Honour, Soft Voice of the Serpent, Six Feet of the Country, Friday's Footprint, Not for Publication.*

Gordon, Caroline: *The Strange Children.*

Grau, Shirley Ann: *The Black Prince.*

Kawabata, Yasunari: *Snow Country, Thousand Cranes.*

Lavin, Mary: *The Great Wave, Happiness, Collected Stories.*

Lehmann, Rosamond: *Dusty Answer, The Echoing Grove, The Ballad and the Source, The Weather in the Streets, Invitation to the Waltz.*

Lessing, Doris: *The Golden Notebook, Children of Violence* (the Martha Quest series of novels).

Litvinov, Ivy: *She Knew She Was Right.*

McCarthy, Mary: *The Company She Keeps, A Charmed Life, The Group, Memories of a Catholic Girlhood* (demi-memoirs, published by Harcourt, Brace, 1957).

McCullers, Carson: *The Member of the Wedding.*

Mauriac, François: *Woman of the Pharisees, Thérèse, The Desert of Love.*

Marshall, Paule: *Brown Girl, Brownstone, The Chosen Place, The Time-less People.*

Moore, Brian: *The Lonely Passion of Judith Hearne, I Am Mary Dunne.*

Mortimer, Penelope: *The Pumpkin Eater, The Home.*

Murdoch, Iris: *Under the Net, The Flight from the Enchanter, The Sand-castle, The Bell, The Red and the Green.*

Nabokov, Vladimir: *Lolita.*

Nin, Anais: *Cities of the Interior,* containing "Ladders to Fire," "Children of the Albatross," "Four-Chambered Heart," "Spy in the House of Love," "Solar Barque."

Oates, Joyce Carol: *A Garden of Earthly Delights.*

O'Connor, Flannery: *Everything That Rises Must Converge, A Good Man Is Hard to Find.*

Olivia (Dorothy Bussy): *Olivia.*

Olsen, Tillie: *Tell Me a Riddle.*

Paley, Grace: *The Little Disturbances of Man.*

Pasternak, Boris: *Doctor Zhivago.*

Pavese, Cesare: *Among Women Only.*

Porter, Katherine Anne: *Flowering Judas, Pale Horse, Pale Rider, The Leaning Tower, Ship of Fools.*

Rao, Raja: *The Serpent and the Rope.*

Rhys, Jean: *Voyage in the Dark, Good Morning, Midnight, Wide Sargasso Sea.*

Richardson, Dorothy: *Pilgrimage* (four volumes, containing thirteen separate novels which form a continuing story).

Rutherford, Dorothea: *The Threshold.*

Salinger, J. D.: *Franny and Zooey.*

Sarraute, Nathalie: *Martureau, Portrait of a Man Unknown, The Plane-tarium, Golden Fruits.*

Slessinger, Tess: *On Being Told That Her Second Husband Has Taken His First Lover and Other Stories.*

Solzhenitsyn, Aleksandr: "Matryona's Home," from *Halfway to the Moon* (London: Weidenfeld & Nicolson, 1964).

Spark, Muriel: *The Girls of Slender Means, The Prime of Miss Jean Brodie, The Mandelbaum Gate.*

Stafford, Jean: *Boston Adventure, The Mountain Lion, The Catherine Wheel, Children Are Bored on Sunday, Bad Characters.*

Stead, Christina: *The Man Who Loved Children, Letty Fox: Her Luck, The Puzzleheaded Girl, Dark Places of the Heart.*

Stein, Gertrude: *Three Lives.*

Taylor, Elizabeth: *A Wreath of Roses.*

Taylor, Peter: *Happy Families Are All Alike, The Widows of Thornton.*

Undset, Sigrid: *Kristin Lavransdatter, The Longest Years, Four Stories, Jenny.*

Vittorini, Elio: *The Dark and the Light.*

von Abele, Rudolph: *The Vigil of Emmeline Gore.*

Waters, Frank: *People of the Valley, The Woman at Otowi Crossing.*

Welty, Eudora: *The Golden Apples, Selected Stories, Delta Wedding, The Bride of the Inisfallen, Losing Battles.*

Wescott, Glenway: *The Grandmothers.*

West, Rebecca: *The Fountain Overflows.*

Williams, William Carlos: *The Farmer's Daughters, White Mule, In the Money, The Build-Up.*

Windham, Donald: *The Warm Country.*

Woolf, Virginia: *The Voyage Out, Night and Day, Mrs. Dalloway, The Waves, To the Lighthouse, Orlando, The Years, Between the Acts,* plus all her essays (collected in four volumes).

A BRIEF LISTING OF PLAYS

The Greek Classics: Sophocles' *Antigone,* Euripides' *Electra* and *Medea,* Aeschylus' *House of Atreus* trilogy.

Shakespeare: *Macbeth, Antony and Cleopatra, Twelfth Night, Much Ado about Nothing, As You Like It, Cymbeline, A Winter's Tale.*

Ibsen, Henrik: *A Doll's House, The Lady from the Sea, The Wild Duck, The Master Builder, Rosmersholm, Hedda Gabler.*

Strindberg, August: *The Father, The Stronger, Easter, The Bridal Crown, There Are Crimes and Crimes, Miss Julie.*

Chekhov, Anton: *The Sea Gull, The Three Sisters, Uncle Vanya, The Cherry Orchard.*

Shaw, Bernard: *Caesar and Cleopatra, Saint Joan, Major Barbara, Pygmalion.*

Synge, John: *Deirdre of the Sorrows.*

Lorca, Federico García: *Yerma, Blood Wedding, House of Bernarda Alba.*

Williams, Tennessee: *A Streetcar Named Desire, The Rose Tattoo.*

POETS

I've limited this to women poets because lyric poetry is the personal voice; when the male poet writes about women, he brings them into *his* universe and cannot see them whole, as they see and write about themselves.

Akhmadulina, Bella: *Fever and other new poems,* translated by Geoffrey Dutton and Igor Mezhakoff-Koriakin. New York: Morrow, 1969.

Akhmatova, Anna: *Selected Poems,* translated by Richard McKane. London: Penguin Books, 1969.

Bergé, Carol: *An American Romance.* Los Angeles: Black Sparrow Press, 1969.

————: *From a Soft Angle: Poems about Women.* Indianapolis: Bobbs-Merrill, 1971.

Bishop, Elizabeth: *Questions of Travel.* New York: Farrar, Straus & Giroux, 1965.

de Vinck, Catherine: *A Time to Gather.* Allendale, N.J.: Alleluia Press, 1967.

Dickinson, Emily: *Selected Poems and Letters,* edited by Robert N. Linscott. Garden City, N.Y.: Doubleday Anchor, 1959.

Doolittle, Hilda (H.D.): *Selected Poems.* New York: Grove Press, 1957. Also her novel *Bid Me to Live.*

Fraser, Kathleen: *Change of Address.* San Francisco: Kayak Books, 1966.

————: *In Defiance of the Rains.* San Francisco: Kayak Books, 1969.

Guest, Barbara: *The Blue Stairs.* New York: Corinth Books, 1968.

Kizer, Carolyn: *The Ungrateful Garden.* Bloomington, Ind.: University of Indiana Press, 1961.

————: *Knock upon the Silence.* Garden City, N.Y.: Doubleday, 1965.

Levertov, Denise: *With Eyes at the Back of Our Heads.* New York: New Directions, 1959.

————: *The Jacob's Ladder.* New York: New Directions, 1961.

————: *O Taste and See.* New York: New Directions, 1965.

————: *The Sorrow Dance.* New York: New Directions, 1966.

————: *Relearning the Alphabet.* New York: New Directions, 1970.

————: *To Stay Alive.* New York: New Directions, 1972.

Moore, Marianne: *Complete Poems.* New York: Macmillan, 1967.

————: *A Marianne Moore Reader.* New York: Viking, 1961.

Piercy, Marge: *Breaking Camp.* Middletown, Conn.: Wesleyan University Press, 1968.

————: *Hard Loving.* Middletown, Conn.: Wesleyan University Press, 1969.

Pitter, Ruth: *Urania.* London: The Cresset Press, 1950.

Plath, Sylvia: *The Colossus.* New York: Knopf, 1962.

————: *Ariel.* New York: Harper & Row, 1966.

————: *The Bell Jar* (a novel). New York: Harper & Row, 1971.

Raine, Kathleen: *Collected Poems.* New York: Random House, 1956.

————: *The Hollow Hill.* London: Hamish Hamilton, 1965.

Randall, Margaret: *October*. Mexico City: El Corno Emplumado, 1965.

————: *Water I Slip into at Night*. Mexico City: El Corno Emplumado, 1967.

————: *25 Stages of My Spine*. New Rochelle: Elizabeth Press, 1967.

Rich, Adrienne: *Necessities of Life*. New York: Norton, 1966.

————: *Leaflets: Poems 1965–1968*. New York: Norton, 1969.

Rukeyser, Muriel: *Waterlily Fire*. New York: Macmillan, 1962.

Sappho: *Poems*, translated by Willis Barnstone. Garden City, N.Y.: Doubleday Anchor, 1965.

————: *Poems*, translated by Mary Barnard. Berkeley: University of California Press, 1958.

Sexton, Anne: *To Bedlam and Part Way Back*. Boston: Houghton Mifflin, 1960.

————: *All My Pretty Ones*. Boston: Houghton Mifflin, 1961.

————: *Live or Die*. Boston: Houghton Mifflin, 1966.

————: *Love Poems*. Boston: Houghton Mifflin, 1969.

Sitwell, Edith: *Collected Poems*. New York: Vanguard, 1954.

Swenson, May: *To Mix With Time*. New York: Scribner's, 1963.

————: *Half Sun Half Sleep*. New York: Scribner's, 1967.

————: *Iconographs*. New York: Scribner's, 1970.

Tsveteyeva, Marina: *Selected Poems*, translated by Elaine Feinstein. Oxford: Oxford University Press, 1971. This is the first volume of translations of this difficult poet.

Valentine, Jean: *Dream Barker and other poems*. New Haven, Conn.: Yale University Press, 1965.

————: *Pilgrims*. New York: Farrar, Straus & Giroux, 1969.

Wakoski, Diane: *Inside the Blood Factory*. Garden City, N.Y.: Doubleday, 1968.

————: *Discrepancies and Apparitions*. Garden City, N.Y.: Doubleday, 1966.

————: *The Magellanic Clouds*. Los Angeles: Black Sparrow Press, 1970.

————: *Greed*, Parts I–IV. Los Angeles: Black Sparrow Press, 1969.

————: *Greed*, Parts V and VI. Los Angeles: Black Sparrow Press, 1971.

————: *The Motorcycle Betrayal Poems*. New York: Simon and Schuster, 1971.

Other women writers, poets, novelists, prose writers, diarists: Edna St. Vincent Millay, Elizabeth Barrett Browning, Dorothy Parker, Christina Rossetti, Bryher, Jane Bowles, Henry Handel Richardson, Nelly Sachs, Edna O'Brien, Lillian Hellman, Shirley Jackson, Rochelle Owens, Mary Austin, Nancy Wilson Ross, Françoise Sagan, Rosalyn Drexler, Susan Sontag, Brigid Brophy, Alison Lurie, Janet Frame, Hortense Calisher,

Zoe Oldenbourg, Sylvia Townsend Warner, Françoise Mallet-Joris, Leonie Adams, Isabel Bolton, Marya Zaturenska, Selma Lagerlöf, Elsa Morante, Elsa Triolet, Elinor Wylie, Rumer Godden, Iris Origo, Emily Hahn, Eleanor Clark, Isabella Gardner, Charlotte Mew, Laura Riding, Rose Macaulay, Pauline Smith, Gertrud von Le Fort, Maria Edgeworth, George Sand, Madame de Staël, Dorothy Wordsworth, Joanne Greenberg, Ann Stanford, Gabriela Mistral, Janet Lewis.

AUTOBIOGRAPHICAL WORKS

Angelou, Maya: *I Know Why the Caged Bird Sings*. New York: Random House, 1969.

Beauvoir, Simone de: *Memoirs of a Dutiful Daughter*. Cleveland: World, 1959.

————: *The Prime of Life*. Cleveland: World, 1962.

————: *Force of Circumstances*. New York: Putnam, 1965.

Bryher: *The Heart to Artemis*. New York: Harcourt, Brace, 1962.

Chagall, Bella: *Burning Lights*. New York: Schocken, 1946.

Day, Dorothy: *The Long Loneliness*. New York: Image Books, 1952.

De Mille, Agnes: *Dance to the Piper*. Boston: Atlantic–Little, Brown, 1952.

————: *And Promenade Home*. Boston: Atlantic–Little, Brown, 1958.

Dunham, Katherine: *A Touch of Innocence*. New York: Harcourt, Brace, 1969.

Hathaway, Katharine Butler: *The Little Locksmith*. New York: Coward-McCann, 1943.

Hurnscot, Loran: *A Prison, A Paradise*. New York: Viking, 1959.

James, Alice: *Diary*, edited by Leon Edel. New York: Dodd, Mead, 1964.

Karsavina, Tamara: *Theatre Street*. New York: Dutton, 1950.

Leduc, Violette: *La Batarde*. New York: Farrar, Straus & Giroux, 1964.

Mallet-Joris, Françoise: *A Letter to Myself*. New York: Farrar, Straus & Giroux, 1964.

Malraux, Clara: *Memoirs*. New York: Farrar, Straus & Giroux, 1967.

Mansfield, Katherine: *Journals*. New York: Knopf, 1946.

————: *Scrapbook*. New York: Knopf, 1940.

————: *Letters to John Middleton Murry*. New York: Knopf, 1951.

Morley, Helena: *Diary of Helena Morley*. New York: Farrar, Straus & Cudahy, 1957.

Nin, Anais: *Diary*, Vols. 1, 2, 3. New York: Harcourt, Brace, 1968, 1969, 1970.

Potter, Beatrix: *Journal*. New York: F. Warne, 1966.

Richards, M. C.: *Centering*. Middletown, Conn.: Wesleyan University Press, 1964.

Sarton, May: *I Knew a Phoenix*. New York: Rinehart, 1959.

————: *Plant Dreaming Deep*. New York: Norton, 1968.

Stark, Freya: *Traveller's Prelude*. London: Penguin Books, 1962.

Schreiner, Olive: *Story of an African Farm*. New York: Fawcett, 1968.

Thomas, Caitlin: *Leftover Life to Kill*. Boston: Atlantic–Little, Brown, 1957.

Woolf, Virginia: *A Writer's Diary*. New York: Harcourt, Brace, 1954.

RELATED IMPORTANT BOOKS

Beauvoir de, Simone: *The Second Sex*. New York: Knopf, 1953.

Blanch, Lesley: *The Wilder Shores of Love*. New York: Simon and Schuster, 1954.

Callahan, Sidney Cornelia: *Illusion of Eve*. New York: Sheed & Ward, 1956.

Dunn, Nell: *Talking to Women*. New York: Ballantine Books, 1968.

Ellmann, Mary: *Thinking about Women*. New York: Harcourt, Brace and World, 1968.

Hardwick, Elizabeth: *A View of My Own*. New York: Farrar, Straus & Giroux, 1962.

Janeway, Elizabeth: *Man's World, Woman's Place*. New York: William Morrow, 1971.

Marcus, Steven: *The Other Victorians*. New York: Basic Books, 1966.

Millett, Kate: *Sexual Politics*. Garden City, N.Y.: Doubleday, 1970.

Woolf, Virginia: *A Room of One's Own*. New York: Harcourt, Brace, 1929.